SpringerBriefs in Geography

SpringerBriefs in Geography presents concise summaries of cutting-edge research and practical applications across the fields of physical, environmental and human geography. It publishes compact refereed monographs under the editorial supervision of an international advisory board with the aim to publish 8 to 12 weeks after acceptance. Volumes are compact, 50 to 125 pages, with a clear focus. The series covers a range of content from professional to academic such as: timely reports of state-of-the art analytical techniques, bridges between new research results, snapshots of hot and/or emerging topics, elaborated thesis, literature reviews, and in-depth case studies.

The scope of the series spans the entire field of geography, with a view to significantly advance research. The character of the series is international and multidisciplinary and will include research areas such as: GIS/cartography, remote sensing, geographical education, geospatial analysis, techniques and modeling, landscape/regional and urban planning, economic geography, housing and the built environment, and quantitative geography. Volumes in this series may analyze past, present and/or future trends, as well as their determinants and consequences. Both solicited and unsolicited manuscripts are considered for publication in this series.

SpringerBriefs in Geography will be of interest to a wide range of individuals with interests in physical, environmental and human geography as well as for researchers from allied disciplines.

More information about this series at http://www.springer.com/series/10050

Wendy Shaw

America's Poorest and Most Affluent Counties, 1980 to 2010

 Springer

Wendy Shaw
Department of Geography and Geographic Information Sciences
Southern Illinois University Edwardsville
Edwardsville, IL, USA

ISSN 2211-4165   ISSN 2211-4173   (electronic)
SpringerBriefs in Geography
ISBN 978-3-030-75339-9   ISBN 978-3-030-75340-5   (eBook)
https://doi.org/10.1007/978-3-030-75340-5

This Springer imprint is published by the registered company Springer Nature Switzerland AG
The registered company address is: Gewerbestrasse 11, 6330 Cham, Switzerland

To my children Shaw Green, Fflur Green, Amanda Schulz (nee Wallace), Kassidy Zhou and my grandchildren Zoe Recklein, Leo Cowley, and Wolfram Schulz. With the hope they and future generations live in a world of increasing equality and inclusion.

Contents

About the Author

Wendy Shaw is a Professor of in the Department of Geography and Geographic Information Sciences at Southern Illinois University Edwardsville. She received her M.S in Geography from the University of Arkansas and a PhD in Geography from the University of Georgia in 1994. Her research interests focus on spatial aspects United States poverty, and the spatial inequality that exists based on characteristics such as race, gender, and educational opportunity.

Chapter 1
Introduction

Abstract The author provides a discussion of the extent of poverty in the United States after Lyndon B. Johnson's declaration of the 'War on Poverty'. This introductory chapter details changes in the number and percentages of the poor at the national level, in order to provide a background for this study. The focus of this study is, in essence, an exploration of the changing spatial distribution of America's poorest and most affluent counties over the 30 years from 1980 to 2010.

Keywords United States · Poverty · War on poverty

Poverty in the United States is a chronic and seemingly intractable problem (Rubinstein 1989; Ropers 1991; Bishaw 2014; Miller and Weber 2014; U.S. Census Bureau 2018), which reemerged into public focus in the early 1960s when President Lyndon B. Johnson declared an "unconditional war on poverty" (Matthews, 2014; Burch 2017). During the 1960s and 1970s poverty was addressed by a host of development and social programs (Mathews 2014) and the topic of poverty engendered a flurry of academic interest (Brunn and Wheeler 1971; Morrill and Wohlenberg 1971). The results of this focus on poverty was dramatic, with the poverty rate falling from 22.2% (39.85 million) in 1960 to 12.6% (25.42 million) just 10 years later in 1970. However, the spotlight on development of disadvantaged regions dimmed as by the mid 1970s it appeared that the war against poverty had been all but won. As is now clear such a celebration of victory was premature. Using the Census Bureau definition of poverty both the percentage (11.1), and the number of people (22.97 million) who were poor, reached a low in 1973. Since 1973 the percentage of people in poverty rose back above the 1970 12.6% level from 1980 through 1998, in 2004, and from 2008 through 2016, reaching 15% or above in 1982, 1983, 1993, 2010, 2011, and 2012 (U.S. Census Bureau 2019). It is disturbing to note that since 2000, when 11.3% or 31.58 million people were officially poor, the poverty rate was on the rise (Eberstadt 1988; Proctor and Dalaker 2003) until 2012. Since 2012 the rate has fallen somewhat; in 2017 12.3% or 39.7 million people were impoverished (Fontenot et al. 2018) which, in terms of sheer numbers, is significantly more than there were in 1965 (33.19 million) during the early years of Johnson's war. By 2019 the poverty rate (10.5%) and the number of impoverished people (33.98 million) continued to fall. However, the slightly less than 34 million poor people in 2019 is a larger number than the impoverished population in 23 of the years between 1969 and 2018 (U.S. Census Bureau 2020). Clearly the number of Americans who fall below the official poverty threshold remains significant and troubling. (Fig. 1.1).

In the years since poverty hit the high point of 15.2% in 1983 (O'Hare 1985), research has often focused on particular segments of the population, including children, female-headed households, African Americans, and the working poor (Jones and Kodras 1990; Eggers and Massey 1991; Hanson and Pratt 1991; Nord and Sheets 1992; Arrighi and Maume 2007; Iceland 2014; Bischoff 2016), as well as on national poverty levels. In addition, there have been some studies that sought to explore the spatial dimensions of U.S. poverty (Brunn and Wheeler 1971; Shaw 1996; Shaefer 2012). Recent studies touching on the geography of U.S. poverty at the county level suggest that it may not only be people in rural counties that are disadvantaged, but also in urban of even suburban counties (Murphy and Allard 2015; Kneebone 2016).

While there has been significant focus on poverty in the United States, including on its spatial characteristics, since the 1960s there has been relatively little research on the concomitant geography of affluence. The geographies of poverty and affluence give a view of spatial economic segregation. The affluent are mobile and have

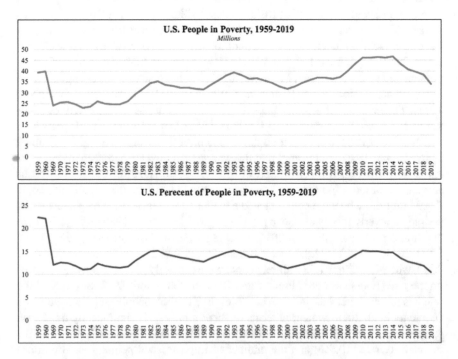

Fig. 1.1 Number of Poor and Percent Poor in the U.S., 1959 to 2017. (Source: U.S. Bureau of the Census, Current Population Survey, Annual Social and Economic Supplements)

a great deal of latitude in choosing where they live. In contrast, the poor have few options and very little mobility (Burns, 1980; Higley 1995; Shaw 1997).

The purpose of this study is to explore the changing spatial distribution of America's poorest and most affluent counties over the 30 years from 1980 to 2010. At the country level the spatial distribution of poverty in 1980 and 1990 was mapped and analyzed by Shaw (1996) and was clearly rural in nature at that spatial scale (Shaw 2000). The geography of affluence for the same period was also discussed by Shaw (1997). The release of the 2000 census data made a preliminary update of the poverty study possible (Shaw 2004), and now the 2010 census data makes it tenable to expand that study and investigate changes in the geography of U.S. poverty since 1980, expressed at the spatial scale of the county. This study is also extended to explore the most affluent counties over the same time period. Spatial aspects of both the poorest and most affluent counties are focused on, as well as the changing gap and relative geographies between rich and poor over three decades.

References

Arrighi, B.A. and D.J. Maume. 2007. *Child Poverty in America today*. Praeger, Westport Connecticut.

Bischoff, K. 2016. *Geography of Economic Inequality. Washington Center for Equitable Growth.* 31 October. https://equitablegrowth.org/geography-of-economic-inequality/ Accessed 29 October 2020.

Bishaw, A. 2014. *Changes in Areas with Concentrated Poverty: 2000 to 2010*. American Community Survey Report ACS-27, June.

Brunn, S.D. and Wheeler, J.O. 1971. Spatial Dimensions of Poverty in the United States. *Geografiska Annaler* 53B (1): 6-15.

Burch, J.R. 2017. *The Great Society and the War on Poverty: An Economic Legacy in Essays and Documents*. Greenwood, 5 June.

Burns, E.K. 1980. The Enduring Affluent Suburb. *Landscape* 24 (1): 33-41.

Eberstadt, N. 1988. Economic and Material Poverty in the U.S. *Public Interest* 90 (Winter): 50-65.

Eggers, M.L. and D.S. Massey. 1991. The Structural Determinants of Urban Poverty: A Comparison of Whites, Blacks, and Hispanics. *Social Science Research* 20 (3): 217-255.

Fontenot, K., J. Semega and M. Kollar. 2018. *Income and Poverty in the United States: 2017*. U.S. Census Bureau, Current Population Report P60–263, September.

Hanson, S. and G. Pratt. 1991. Job Search and Occupational Segregation of Women. *Annals of the Association of American Geographers* 81 (2): 229-253.

Higley, S.R. 1995. *Privilege, Power, and Place: The Geography of the American Upper Class*. Lanham MD: Rowman and Littlefield.

Iceland, J. 2014. *A Portrait of America: The Demographic Perspective*. University of California Press, Oakland.

Jones, J.P. and J.E. Kodras. 1990. Restructured Regions and Families: The Feminization of Poverty in the United States. *Annals of the Association of American Geographers* 80 (2): 163-183.

Kneebone, E. 2016. *Urban and Suburban Poverty: The Changing Geography of Disadvantage*. Penn Institute for Urban Research, 10 February. https://penniur.upenn.edu/publications/urban-and-suburban-poverty-the-changing-geography-of-disadvantage Accessed 31 October 2020.

Matthews, D. 2014. *Everything You Need to Know about the War on Poverty*. https://www.washingtonpost.com/news/wonk/wp/2014/01/08/everything-you-need-to-know-about-the-war-on-poverty/ 8 January. Accessed 5 November 2020.

Miller, K. and Weber, B. 2014. *Persistent Poverty Dynamics: Understanding Poverty Trends over 50 Years*. Rural Poverty Research Institute, July.

Morrill, R.L. and E.H. Wohlenberg. 1971. *The Geography of Poverty in the United States*. McGraw-Hill Book Co., Problems Series.

Murphy, A.K. and S.W. Allard. 2015. The Changing Geography of Poverty. *Focus* 32 (1): 19–23, Spring/Summer.

Nord, S. and R.G. Sheets. 1992. *Service Industries and the Working Poor in Major Metropolitan Areas in the United States*. In Sources of Metropolitan Growth. E.S. Mills and J.F. McDonald, Eds., Center for Urban Policy Research, New Brunswick, N.J.

O'Hare, W.P. 1985. Poverty in America: Trends and New Patterns. *Population Bulletin* 40 (June): 2-43.

Proctor, B.D. and J. Dalaker. 2003. *Poverty in the United States: 2002*. U.S. Census Bureau, Current Population Report, P60–222, Sept. 33p.

Ropers, R.H. 1991. Persistent Poverty. *The American Dream Turned Nightmare*. Insight Books, Plenum Press, N.Y.

Rubinstein, E. 1989. Losing More Ground. *National Review* 41 (May):13.

Shaefer, H.L. 2012. *Extreme Poverty in the United States, 1996–2011*. National Poverty Center, Policy Brief 28, February.

Shaw, W. 1996. *The Geography of United States Poverty. Patterns of Deprivation, 1980–1990*. Garland Publishing Inc., New York, Fall.

Shaw, W. 1997. The Spatial Concentration of Affluence in the United States. The Geographical Review, 87 (4): 546-553

Shaw, W. 2000. Illinois: Spatial Scales of Poverty. *The Geographical Bulletin*, 47 (1): 9–22, May.

Shaw, W. 2004. The Changing Location and Characteristics of America's Poorest Counties. *Proceedings of the Applied Geography Conference*, 27:448–457, Oct.

U.S. Census Bureau. 2018. *Income, Poverty and Health Insurance Coverage in the United States: 2017*. https://www.census.gov/newsroom/press-releases/2018/income-poverty.html Accessed 1 November 2020.

U.S. Census Bureau. 2019. *Historical Poverty Tables: People and Families – 1959 to 2017*. https://www.census.gov/data/tables/time-series/demo/income-poverty/historical-poverty-people.html Accessed 4 November 2020

U.S. Census Bureau. 2020. *Income and Poverty in the United States: 2019* https://www.census.gov/library/publications/2020/demo/p60-270.html Accessed 5 November 2020

Chapter 2
Data and Methodology

Abstract A general measure of poverty and affluence (INDEX) was developed and is explained. INDEX differs from other commonly used measures as it is sensitive to geographic variation in the cost of living. Information concerning the sources of the data used are provided, and the methodologies employed in the study are outlined.

Keywords Measuring poverty · Methodology · Data

2.1 Data and Methodology

In order to investigate the spatial distribution of the poorest and most affluent counties in the United States, an overall measure of poverty and affluence must be selected. Traditional measures such as percent of the population below the official poverty line, per capita income, and median household income lack sensitivity to the substantial spatial variation in the cost of living. A simple but spatially sensitive measure INDEX was developed in a previous 1980 and 1990 analysis (Shaw 1996). Since lack of income is still seen as fundamental to poverty in the United States an income measure is appropriate, but it must incorporate some spatial sensitivity to variations in the cost of living.

It is housing costs that vary the most spatially, so these costs were adjusted for by developing a 'INDEX' for each county as follows:

$$\frac{\text{Median Household Income} - \text{Median Housing Costs}}{\text{Average number of persons per household}}$$

Essentially INDEX as calculated is per capita income that remains after housing costs are accounted for.

The necessary data to calculate INDEX have been stably available since the 1980 census (U.S. Department of Commerce 1980, 1990, 2000, 2010) and so are used for this extended study. Incorporating median housing costs (which include mortgage payments, real estate taxes, insurance costs, and utilities), into INDEX provides a

© The Author(s), under exclusive license to Springer Nature
Switzerland AG 2021
W. Shaw, *America's Poorest and Most Affluent Counties, 1980 to 2010*,
SpringerBriefs in Geography, https://doi.org/10.1007/978-3-030-75340-5_2

5

measure of economic well-being that is sensitive to spatial variations in the cost of living, unlike historic census data using the official poverty line. This income INDEX for U.S. counties was mapped for 1980, 1990, 2000, and 2010 to explore the spatial distribution of poor and affluent counties. In addition, a statistical analysis of the 1980–1990 data is presented, simple descriptive statistics are utilized as a starting point to gain some understanding of changes in the geography of poverty and affluence over time, and selected case study counties are presented to provide more in-depth context.

References

Shaw, W. 1996. *The Geography of United States Poverty. Patterns of Deprivation, 1980–1990.* Garland Publishing Inc., New York, Fall.

U.S. Department of Commerce. 1980. *Census of the Population.* U.S. Bureau of the Census. http://www.census.gov/ Accessed October 2020

U.S. Department of Commerce. 1990. *Census of the Population.* U.S. Bureau of the Census. http://www.census.gov/ Accessed October 2020

U.S. Department of Commerce. 2000. *Census of the Population.* U.S. Bureau of the Census. http://www.census.gov/ Accessed October 2020

U.S. Department of Commerce. 2010. *American Community Survey.* https://www.census.gov/programs-surveys/acs/data.html Accessed October 2020.

Chapter 3
The Geography of Poverty in United States at the Spatial Scale of the County

Abstract This chapter includes identification of the poorest counties in 1980, 1990, 2000, and 2010, and details about cores of persistent poverty. In addition the author discusses changes in poverty over the 30-year study period. Tables that detail the poorest 50, 100, and 157 (5%) of counties are provided for each study year. Maps of the poorest 5% of counties for 1980, 1990, 2000, and 2010 are also included in the chapter.

Keywords Poorest counties · Poverty cores · 1980 · 2010 · Maps

3.1 U.S. Poverty in 1980

The poorest 5% of counties in terms of INDEX for 1980 are presented in Fig. 3.1 which reveals that a distinct spatial pattern characterized U.S. poverty. Five core areas of poverty existed in 1980. The first poverty core was the 'Western' core, located in the region encompassed by southern Utah, northeast Arizona, western New Mexico, and south-central Colorado. This poverty region was located in an area with relatively high (on average 26%) Native American population (U.S. Department of Commerce 1983). Indeed, much of this poverty core, including the intensely impoverished Apache County, Arizona, is located on reservation land.

A second area of intense poverty is the 'Dakotas' core, which existed throughout central South Dakota extending into south-central North Dakota. This poverty core has a very high (on average over 60%) Native American population (U.S. Department of Commerce 1983). Again, much of this poverty core, such as Corson, Shannon, Todd, and Ziebach Counties is reservation land.

A third distinct poverty area, 'Texas Border' core, extended almost the entire length of the Texas-Mexico border. This is in keeping with the research of Jones and Kodras who link poverty status to Hispanic ethnicity (Jones and Kodras 1990). Of the 13 Texan counties with the highest percentage of Hispanic population, 11 were contained within the identified poverty core. On average the counties in this core were over 67% Hispanic-Latino, and the most intensely impoverished counties of Maverick, Presido, Starr, and Zavala had approximately 92% Hispanic-Latino

W. Shaw, *America's Poorest and Most Affluent Counties, 1980 to 2010*,
SpringerBriefs in Geography, https://doi.org/10.1007/978-3-030-75340-5_3

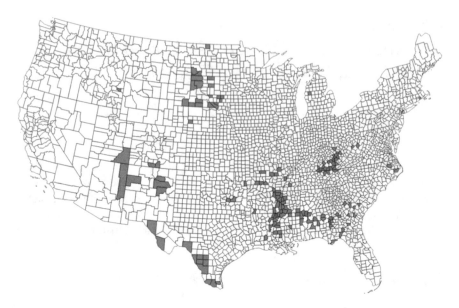

Fig. 3.1 The poorest 5% of counties, 1980

population compared with 21% for the state as a whole (U.S. Department of Commerce 1983).

A fourth broad area of poverty, the 'Southern' core can be seen along the Mississippi River from Louisiana to southern Illinois; this poverty region also spanned much of Mississippi as well as the southern coastal plain from Alabama to North Carolina. The region identified supports the comments of other researchers (Brunn and Wheeler 1971; Smith 1972, 1990; Jones and Kodras 1990) who suggest that the concentration of poverty across the South mirrors the distribution of the rural Black population. The most intensely impoverished of the counties contained within this core lay along the Mississippi River from central Arkansas and northern Mississippi to southern Mississippi and northern Louisiana, and in southern Alabama. All these counties, with the exception of West Carroll County, Louisiana, had substantially higher percentages (on average over 54) of Black population than the averages for their state. In these most severely impoverished counties in 1980 Black residents comprised a majority of the population (U.S. Department of Commerce 1983).

A fifth 'Appalachian' core of poverty, which was particularly geographically coherent, spanned most of eastern Kentucky and extended into northern Tennessee. In contrast to the four previously identified cores of poverty this area had no concentration of minority population. Instead, the population of the counties in this region are, on average, over 98% White non-Hispanic Latino.

The concentration of poverty in 1980 in relatively few core areas is further illustrated by Table 3.1. The poorest 50 counties are located in just 11 states; moreover, these same 11 states account for 124 of the poorest 150 counties. Two states alone

(Mississippi and Kentucky) encompass 20 of the poorest 50, 34 of the poorest 100, and 50 of the poorest 157 (5%) of counties. The poorest 5% of counties were located in a total of 24 states in 1980. It is the poverty areas centered in Mississippi and Kentucky that can be said to represent the cores of the most intense poverty within the United States. In 1980 the poorest county in the United States was Tunica County, Mississippi, a county located in the northwest of the state that borders the Mississippi river (Fig. 3.2).

Table 3.1 Distribution of the Poorest Counties by State: 1980, [1990], (2000), and 2010*

State	In 50 poorest counties	In 51–100 poorest counties	In 101–157 poorest counties	Total in poorest 5% of counties
Kentucky	11 [12] (9) 10*	8 [9] (14) 5*	6 [8] (8) 8*	25 [29] (31) 23*
Mississippi	9 [11] (8) 10*	6 [6] (9) 7*	10 [7] (4) 7*	25 [24] (21) 24*
South Dakota	6 [3] (5) 4*	6 [5] (3) 1*	4 [0] (1) 2*	16 [8] (9) 7*
Georgia	2 [0] (0) 2*	7 [2] (3) 8*	6 [3] (8) 12*	15 [5] (11) 22*
Alabama	6 [3] (5) 2*	2 [2] (2) 5*	3 [4] (3) 4*	11 [9] (10) 11*
Arkansas	4 [3] (0) 1*	4 [3] (3) 2*	2 [4] (5) 1*	10 [10] (8) 4*
Texas	5 [9] (11) 12*	4 [5] (5) 2*	1 [5] (1) 4*	10 [19] (17) 18*
Louisiana	3 [4] (4) 0*	2 [4] (3) 3*	3 [11] (5) 1*	8 [19] (12) 4*
Tennessee	1 [0] (0) 1*	2 [1] (1) 2*	3 [2] (2) 4*	6 [3] (3) 7*
Missouri	0 [0] (0) 0*	2 [1] (0) 1*	3 [2] 2) 1*	5 [3] (2) 2*
New Mexico	0 [3] (1) 0*	4 [0] (2) 2*	0 [1] (2) 0*	4 [4] (5) 2*
North Dakota	0 [0] (1) 0*	1 [1] (0) 3*	2 [1] (0) 0*	3 [2] (1) 3*
Oklahoma	0 [0] (0) 0*	0 [1] (0) 0*	3 [1] (0) 2*	3 [2] (0) 2*
Colorado	2 [0] (1) 1*	0 [3] (0) 0*	0 [1] (1) 0*	2 [4] (2) 1*
Florida	0 [0] (0) 0*	0 [0] (0) 0*	2 [0] (0) 1*	2 [0] (0) 1*
North Carolina	0 [0] (0) 0*	0 [0] (0) 2*	2 [0] (0) 4*	2 [0] (0) 6*
West Virginia	0 [0] (1) 0*	0 [5] (2) 1*	2 [2] (5) 0*	2 [7] (8) 1*
Arizona	1 [1] (1) 0*	0 [0] (0) 1*	0 [2] (2) 1*	1 [3] (3) 2*
Nebraska	0 [0] (0) 0*	1 [0] (0) 0*	0 [0] (0) 0*	1 [0] (0) 0*
New York	0 [0] (3) 2*	1 [0] (0) 0*	0 [1] (0) 0*	1 [1] (3) 2*
Idaho	0 [0] (0) 0*	0 [0] (0) 1*	1 [0] (1) 0*	1 [0] (1) 1*
Michigan	0 [0] (0) 0*	0 [0] (0) 0*	1 [0] (0) 0*	1 [0] (0) 0*
South Carolina	0 [0] (0) 3*	0 [0] (1) 1*	1 [0] (1) 1*	1 [0] (2) 5*
Utah	0 [0] (0) 0*	0 [1] (1) 0*	1 [0] (0) 0*	1 [1] (1) 0*
Virginia	0 [0] (0) 1*	0 [1] (0) 2*	0 [1] (2) 1*	0 [2] (2) 4*
Wisconsin	0 [1] (0) 1*	0 [0] (0) 0*	0 [0] (1) 0*	0 [1] (1) 1*
Montana	0 [0] (0) 0*	0 [0] (0) 0*	0 [0] (2) 0*	0 [0] (2) 0*
California	0 [0] (0) 0*	0 [0] (1) 1*	0 [0] (0) 2*	0 [0] (1) 3*
Ohio	0 [0] (0) 0*	0 [0] (0) 0*	0 [0] (0) 1*	0 [0] (0) 1*

Fig. 3.2 Mississippi counties (Tunica county)

3.2 U.S. Poverty in 1990

The 1990 ranking by INDEX of counties reveals a relative stability within the poorest segment of counties over the 1980s (Fig. 3.3). Of the 157 counties identified as within the poorest 5% in 1980, 104 still remained in this segment in 1990. While some counties, such as Faulk, Miner, and Sanborn Counties, South Dakota, did improve significantly and no longer appeared in the poorest group of counties, most remained among the poorest counties in America. For those counties that remained within the most impoverished 5% there was some variation in relative position. Starr County, Texas (Fig. 3.4) took over the dubious honor of being the most impoverished county in 1990 from Tunica County, Mississippi. Tunica County, meanwhile rose seven positions in the ranking. Such changes do little to change the spatial picture of American poverty in 1990, and the same states embrace the largest number of impoverished counties as in 1980 (Table 3.1). By 1990 the poorest 50 counties are located in just 9 states; these same 9 states account for 107 of the poorest 150 counties. Two states alone (Mississippi and Kentucky) encompass 23 of the poorest 50, 38 of the poorest 100, and 53 of the poorest 157 (5%) of counties. The map of the poorest 5% of counties in 1990 (Fig. 3.3) shows that these counties are even more spatially concentrated than they were in 1980. Coherent areas of extreme poverty clearly remain centered in South Dakota, eastern Arizona, the Texas-Mexican border, the southern Mississippi River valley-southern coastal plain, and the central Appalachians.

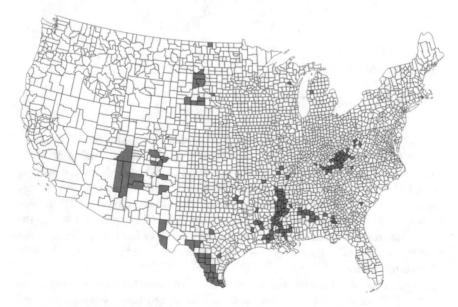

Fig. 3.3 The poorest 5% of counties, 1990

Fig. 3.4 Texas counties (Starr county)

3.3 U.S. Poverty in 2000

The spatial distribution of the poorest 5% of counties in 2000 is presented in Fig. 3.5. Clearly the general spatial pattern has continued to be relatively stable, with the same five cores of poverty identified in both 1980 and 1990 readily apparent. Indeed, of the 157 counties included in the poorest 5% in 1990, 114 remained in this poorest segment in 2000. An additional 15 counties that were among the poorest 5% in 1980 but not in 1990 reentered this poorest segment in 2000. Once again counties in the same states dominate the listing of the poorest counties (Table 3.1). Forming the most temporally stable core of poverty, 88 counties remained in the poorest 5% in all three study years.

Despite the general spatial stability of U. S. poverty over the 20 year period from 1980 to 2000, there are some changes to note. Interestingly the county characterized

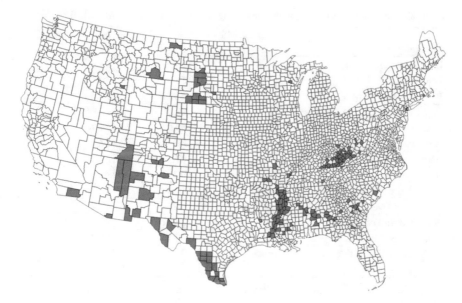

Fig. 3.5 The poorest 5% of counties, 2000

Fig. 3.6 New York counties (New York county)

as the poorest in the United States once more shifted. In 1980 it was Tunica County, Mississippi (ranked 8th poorest in 1990 and 99th poorest in 2000); in 1990 it was Starr County, Texas (ranked 3rd poorest in both 1980 and 2000). In 2000 the poorest county was New York County, New York (ranked 189th poorest in 1990 and 249th poorest in 1980) (Fig. 3.6). For the first time within the study period the poorest county was clearly urban. New York County, New York is the quintessentially urban county with a very different demographic, social, and economic profile than Tunica County, Mississippi and Starr County Texas. Table 3.2 provides some basic information in order to illustrate this contrast.

Tunica County, Mississippi, the most impoverished county in 1980, can clearly be characterized as rural. Population density is low, farm sizes are large, and the people

Table 3.2 Characteristics of the county considered the poorest in 1980, 1990, and 2000

Characteristic	Tunica, MS (*1980*)	Starr, TX (*1990*)	New York, NY (*2000*)
Land Area (Square Miles)	455	1223	23
Population, 2000	9227	53,597	1,537,195
Pop. Change, 1990–2000 (%)	+13.0	+32.3	+3.3
Pop. Density (Sq. Mile 1990)	17	43.8	66,940.1
White (%)	27.5	87.9	54.4
Black (%)	70.2	0.1	17.4
Hispanic (%)	2.5	97.5	27.2
Asian (%)	0.4	0.3	9.4
Aged Under 5 (%)	8.4	10.4	4.9
Aged 65 or Over (%)	10.1	8.2	12.2
Foreign Born Persons (%)	0.9	36.9	29.4
Language Not English (%)*	2.4	90.7	41.9
High School Graduation (%)	60.5	34.7	78.7
Bachelor's Degree (%)	9.1	6.9	49.4
Home Ownership Rate (%)	51.7	79.5	20.1
Median Rent ($)	99	163	478
Median House Value ($)	35,400	21,900	487,300
Unemployment Rate (%)	9.3	20.9	7.1
Median Household Income ($)	23,270	16,504	47,030
Per Capita Money Income ($)	11,978	7069	42,922
Number of Farms	162	885	0
Average Farm Size (acres)	1514	677	0
Metropolitan Areas	None	None	New York, NY

Source: http://www.census.gov/quickfacts Accessed 13 February 2004

of Tunica County are primarily Black. Starr County, Texas poorest in 1990, is also rural but both the total population and population density are higher than in Tunica County. In addition, the population in Starr County is overwhelmingly Hispanic-Latino, over one third are foreign born, and over 90% do not speak English at home (but can be assumed to speak primarily Spanish). High school graduation rates are also substantially lower in Starr County than they are in Tunica County. While Tunica and Starr counties clearly have both differences and similarities when compared to each other, New York County, New York, which was the poorest county in 2000, has virtually no commonality with either Tunica County or Starr County. New York County is within the New York Primary Metropolitan Statistical Area (PSMA) and extends over a very small area, yet is home to over 1.5 million people. With a population density approaching sixty seven thousand per square mile, and racial and ethnic diversity not seen in Tunica or Starr counties, New York County is clearly urban in every sense. The residents in New York County are well educated with almost half the population holding a bachelor's degree, but they must deal with extremely high property values. Home ownership rates in both Tunica and Starr counties are much higher than in New York County. The unemployment rate in New York County is low especially compared to that in Starr County and income relatively high, yet when housing costs are taken into consideration it is the quintessentially urban county of New York, New York that can be considered the poorest county in 2000 when using INDEX as a measure. Unlike the definition of poverty used by the U.S. Census Bureau, INDEX incorporates housing costs, and it is this consideration of housing costs that is responsible for identifying New York County, New York as poor.

3.4 U.S. Poverty in 2010

Figure 3.7 displays the spatial distribution of the poorest 5% of counties in 2010. The general spatial pattern characterized by five cores of poverty has persisted in the 30 years since 1980. Of the poorest 50 counties in 2010, 10 are in Kentucky, 10 in Mississippi, and 12 are in Texas (Table 3.1). Of the 157 counties included in the poorest 5% in 2010, 24 are in Mississippi, 23 in Kentucky, 22 in Georgia, 18 in Texas, and 11 in Alabama. This poorest 5% of counties are distributed across 29 states and so might appear to be relatively geographically dispersed. However, the five cores of poverty identified cross state lines and are visually quite clear when the counties are mapped (Fig. 3.7).

Over the entire study period 17 counties remained in the poorest 50 in 1980, 1990, 2000, and 2010 and another 16 counties remained in the poorest 50 for three of those years. These 33 counties are located in just 8 states, and only 4 states had the 17 counties in the poorest 50 in all four years: Kentucky (5), Texas (4), Mississippi (3), South Dakota (3), Alabama (2) (Table 3.3). Mapping the counties

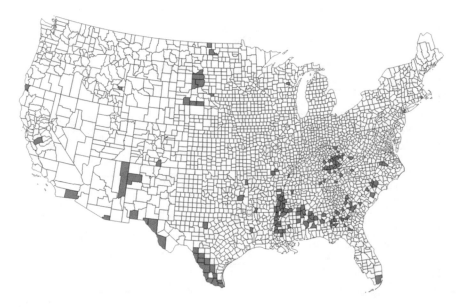

Fig. 3.7 The poorest 5% of counties, 2010

Table 3.3 Counties among the poorest 50 in at least three years; 1980, 1990, 2000 and 2010

County	In poorest 50 in 1980	In poorest 50 in 1990	In poorest 50 in 2000	In poorest 50 in 2010
Greene County, Alabama	Yes	Yes	Yes	Yes
Wilcox County, Alabama	Yes	Yes	Yes	Yes
Apache County, Arizona	Yes	Yes	Yes	No
Chicot County, Arkansas	Yes	Yes	No	Yes
Breathitt County, Kentucky	Yes	Yes	Yes	Yes
Clay County, Kentucky	Yes	Yes	Yes	Yes
Jackson County, Kentucky	Yes	Yes	No	Yes
Knox County, Kentucky	Yes	Yes	Yes	Yes
McCreary County, Kentucky	Yes	Yes	No	Yes
Magoffin County, Kentucky	Yes	Yes	Yes	Yes
Martin County, Kentucky	No	Yes	Yes	Yes
Owsley County, Kentucky	Yes	Yes	Yes	Yes
Wolfe County, Kentucky	Yes	Yes	No	Yes
East Carroll Parish, Louisiana	Yes	Yes	Yes	No
Madison Parish, Louisiana	Yes	Yes	Yes	No
Bolivar County, Mississippi	Yes	Yes	No	Yes
Coahoma County, Mississippi	Yes	Yes	Yes	Yes
Holmes County, Mississippi	Yes	Yes	Yes	Yes
Humphreys County, Mississippi	Yes	Yes	Yes	Yes
Issaquena County, Mississippi	No	Yes	Yes	Yes
Jefferson County, Mississippi	Yes	Yes	Yes	No
Sharkey County, Mississippi	Yes	Yes	Yes	No
Buffalo County, South Dakota	Yes	Yes	Yes	Yes
Shannon County, South Dakota	Yes	Yes	Yes	Yes

(continued)

Table 3.3 (continued)

County	In poorest 50 in 1980	In poorest 50 in 1990	In poorest 50 in 2000	In poorest 50 in 2010
Todd County, South Dakota	Yes	Yes	Yes	Yes
Dimmit County, Texas	No	Yes	Yes	Yes
Hidalgo County, Texas	No	Yes	Yes	Yes
Maverick County, Texas	Yes	Yes	Yes	Yes
Presidio County, Texas	Yes	Yes	Yes	No
Starr County, Texas	Yes	Yes	Yes	Yes
Webb County, Texas	Yes	Yes	Yes	Yes
Willacy County, Texas	No	Yes	Yes	Yes
Zavala County, Texas	Yes	Yes	Yes	Yes

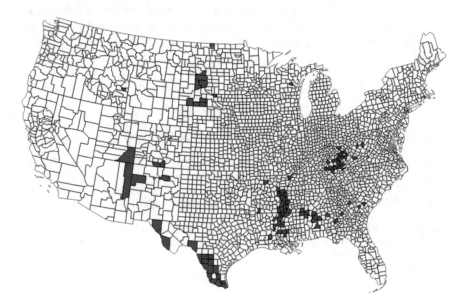

Fig. 3.8 Counties in the poorest 5% in at least three study years, 1980, 1990, 2000, and 2010

that were in the poorest 5% for at least three of the four study years clearly reveals the same five cores of poverty that have been discussed previously (Fig. 3.8).

In 2000 the poorest county was urban, and in 2010 this was again the case. By 2010 rather than New York County being the poorest, it was neighboring Bronx County (Fig. 3.9). By 2010 poverty in the United States, viewed at the spatial scale of the county, remains largely rural. However, the poorest county in 2010 was urban primarily because of high housing costs, as was the case in 2000. Interesting New York County was not even close to being included amongst the poorest counties by 2010. It seems likely that because of the limited spatial extent of these urban

Fig. 3.9 New York counties (The Bronx)

counties, people of limited income are somewhat mobile and move to adjacent areas in order to lower their housing costs. Looking across the study period it is actually the Bronx and Kings County that are the New York counties included most consistently in the impoverished grouping.

References

Brunn, S.D. and Wheeler, J.O. 1971. Spatial Dimensions of Poverty in the United States. *Geografiska Annaler* 53B (1): 6-15.

Jones, J.P. and J.E. Kodras. 1990. Restructured Regions and Families: The Feminization of Poverty in the United States. *Annals of the Association of American Geographers* 80 (2): 163-183.

Smith, D.M. 1972. Towards a Geography of Social Well-being: Inter-state Variations in the United States. *Antipode* Monographs in Social Geography 1, Geographical Perspectives on American Poverty: 5–16.

Smith, L. 1990. The Face of Rural Poverty. *Fortune* (Dec 31st): 100–11.

U.S. Department of Commerce. 1983. *County and City Data Book*. U.S. Bureau of the Census.

Chapter 4
The Geography of Affluence in the United States at the Spatial Scale of the County

Abstract The most affluent counties in 1980, 1990, 2000, and 2010 are identified. This chapter includes details concerning the spatial aspects of affluence, and the changes in affluence over the 30-year study period. Tables that detail the most affluent 50, 100, and 157 (5%) of counties are provided for each study year. The most affluent 5% of counties are mapped for 1980, 1990, 2000, and 2010.

Keywords Affluent counties · 1980 · 2010 · Maps

4.1 U.S. Affluence in 1980

The most affluent 5% of counties in terms of INDEX for 1980 are presented in Fig. 4.1 and the spatial distribution of affluence in 1980 is further illustrated by Table 4.1. The most affluent 50 counties are present in 20 states, which is a significantly larger number than the 11 states that are the location of the poorest 50 counties. While the most affluent counties are not as clustered in particular states as the poorest, the greatest number of the 50 most affluent counties are in Virginia (6), California (4), Colorado (4), and Texas (4). Interesting three states, Colorado, Georgia, and Texas are the location of counties in both the poorest and most affluent 50. Looking at the most affluent 5% of counties compared to the poorest 5%, they are located in 28 states compared to 24 for the poorest counties. Ten states have counties in both the most affluent and poorest 5%. Four states have 10 or more counties within the most affluent 5%; Illinois (16), Virginia (14), Texas (13), and Indiana (10). The most affluent county in 1980 was Los Alamos County, New Mexico. Los Alamos County represents an unusual situation, and has an INDEX that is over 226% higher than the next most affluent county in New Mexico. Los Alamos is very small compared to other counties in New Mexico with an area of approximately 109 square miles, and a population of just over 18,000 in 1990 (Fig. 4.3). The county has a unique history, and was created in 1949 out of portions of adjacent Sandoval and Santa Fe counties that had been the location for the Manhattan Project from 1942 to 1946. The area was administered by the federal government during the project, and the legacy from the specialized purpose for

W. Shaw, *America's Poorest and Most Affluent Counties, 1980 to 2010*,
SpringerBriefs in Geography, https://doi.org/10.1007/978-3-030-75340-5_4

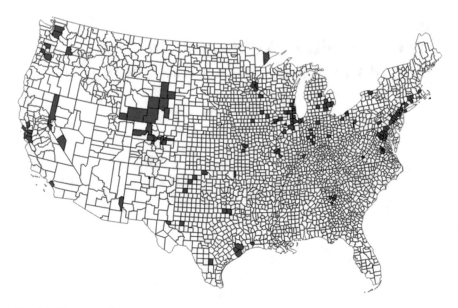

Fig. 4.1 The most affluent 5% of counties, 1980

Table 4.1 Distribution of the most affluent counties by state: 1980, [1990], (2000), and 2010*

State	In 50 most affluent counties	In 51–100 most affluent counties	In 101–157 most affluent counties	Total in most affluent 5% of counties
Virginia	6 [9] (11) 13*	7 [10] (4) 8*	1 [4] (11) 5*	14 [23] (26) 26*
California	4 [2] (2) 0*	1 [3] (2) 1*	2 [1] (1) 0*	7 [6] (5) 1*
Colorado	4 [1] (2) 2*	2 [2] (5) 7*	3 [3] (6) 5*	9 [6] (13) 14*
Texas	4 [1] (2) 3*	5 [2] (3) 4*	4 [1] (2) 6*	13 [4] (7) 13*
Illinois	3 [3] (1) 0*	6 [3] (5) 2*	7 [3] (2) 4*	16 [9] (8) 6*
Indiana	3 [1] (1) 1*	2 [3] (2) 1*	5 [3] (3) 2*	10 [7] (6) 4*
Maryland	3 [5] (4) 9*	2 [6] (5) 1*	2 [0] (2) 1*	7 [11] (11) 11*
Missouri	3 [1] (1) 0*	1 [2] (2) 2*	0 [1] (1) 0*	4 [4] (4) 2*
New Jersey	3 [6] (4) 4*	2 [3] (4) 2*	1 [3] (2) 1*	6 [12] (10) 7*
Wyoming	3 [0] (0) 2*	2 [0] (1) 1*	3 [0] (0) 3*	8 [0] (1) 6*
Michigan	2 [2] (2) 0*	3 [1] (1) 1*	2 [3] (4) 1*	7 [6] (7) 2*
Minnesota	2 [1] (4) 2*	3 [4] (3) 2*	2 [2] (1) 4*	7 [7] (8) 8*
Washington	2 [0] (0) 0*	1 [1] (0) 0*	1 [1] (1) 0*	4 [2] (1) 0*
Wisconsin	2 [1] (2) 2*	0 [1] (1) 0*	2 [1] (2) 0*	4 [3] (5) 2*
Georgia	1 [3] (2) 2*	1 [0] (3) 1*	2 [4] (2) 2*	4 [7] (7) 5*
Kansas	1 [1] (1) 1*	0 [0] (0) 1*	0 [0] (0) 1*	1 [1] (1) 3*
New Mexico	1 [1] (1) 1*	0 [0] (0) 0*	0 [0] (0) 0*	1 [1] (1) 1*
New York	1 [4] (2) 1*	2 [3] (0) 1*	1 [2] (2) 1*	4 [9] (4) 3*
Pennsylvania	1 [2] (2) 1*	0 [1] (1) 2*	6 [2] (0) 1*	7 [5] (3) 4*
West Virginia	1 [0] (0) 0*	0 [0] (0) 0*	1 [0] (0) 0*	2 [0] (0) 0*

(continued)

Table 4.1 (continued)

State	In 50 most affluent counties	In 51–100 most affluent counties	In 101–157 most affluent counties	Total in most affluent 5% of counties
Iowa	0 [0] (0) 0*	2 [0] (0) 0*	2 [1] (1) 2*	4 [1] (1) 2*
Massachusetts	0 [2] (1) 1*	2 [1] (1) 1*	0 [1] (1) 1*	2 [4] (3) 3*
Ohio	0 [0] (1) 1*	2 [0] (2) 1*	2 [6] (1) 2*	4 [6] (4) 4*
Connecticut	0 [4] (2) 2*	1 [1] (2) 1*	3 [2] (1) 1*	4 [7] (5) 4*
Nevada	0 [0] (0) 0*	1 [0] (0) 3*	2 [0] (1) 1*	3 [0] (1) 4*
Oklahoma	0 [0] (0) 0*	1 [0] (0) 0*	1 [0] (0) 0*	2 [0] (0) 0*
Oregon	0 [0] (0) 0*	1 [0] (0) 0*	1 [2] (0) 0*	2 [2] (0) 0*
Arizona	0 [0] (0) 0*	0 [0] (0) 0*	1 [0] (0) 0*	1 [0] (0) 0*
Delaware	0 [0] (0) 0*	0 [1] (0) 0*	0 [0] (1) 0*	0 [1] (1) 0*
New Hampshire	0 [0] (0) 0*	0 [1] (1) 1*	0 [1] (1) 0*	0 [2] (2) 1*
Tennessee	0 [0] (1) 1*	0 [1] (0) 0*	0 [0] (0) 0*	0 [1] (1) 1*
Alabama	0 [0] (0) 0*	0 [0] (1) 1*	0 [2] (0) 0*	0 [2] (1) 1*
Florida	0 [0] (0) 0*	0 [0] (0) 0*	0 [2] (1) 0*	0 [2] (1) 0*
North Carolina	0 [0] (0) 0*	0 [0] (1) 0*	0 [2] (1) 0*	0 [2] (2) 0*
Rhode Island	0 [0] (0) 0*	0 [0] (0) 1*	0 [2] (2) 1*	0 [2] (2) 2*
Kentucky	0 [0] (1) 0*	0 [0] (0) 1*	0 [1] (2) 1*	0 [1] (3) 2*
Vermont	0 [0] (0) 0*	0 [0] (0) 0*	0 [1] (0) 0*	0 [1] (0) 0*
Nebraska	0 [0] (0) 0*	0 [0] (0) 0*	0 [0] (1) 1*	0 [0] (1) 1*
Utah	0 [0] (0) 0*	0 [0] (0) 1*	0 [0] (1) 0*	0 [0] (1) 1*
North Dakota	0 [0] (0) 1*	0 [0] (0) 2*	0 [0] (0) 4*	0 [0] (0) 7*
South Dakota	0 [0] (0) 0*	0 [0] (0) 0*	0 [0] (0) 3*	0 [0] (0) 3*
Montana	0 [0] (0) 0*	0 [0] (0) 0*	0 [0] (0) 2*	0 [0] (0) 2*
Louisiana	0 [0] (0) 0*	0 [0] (0) 0*	0 [0] (0) 1*	0 [0] (0) 1*

which the area was delineated is still apparent today. For example, 65.5% of the population 25 or older hold a bachelor's degree or higher, and median household income is $110,190 (Encyclopedia Britannica 2021; U.S. Census Bureau 2018a; Wikipedia 2021b).

The spatial distribution of these counties lends support to the contention that suburban counties associated with metropolitan areas tend to be affluent. While not all metropolitan areas include counties in the most affluent 5%, many do. Of these most affluent 157 counties, 126 (over 80%) are located in 49 metropolitan areas, with several metropolitan areas having multiple such counties. The most affluent 50 counties in 1980 are dominated by Washington DC (7), Chicago (4), New York (4), and San Francisco, with (4) (Table 4.2). One particularly apparent exception to this metropolitan spatial pattern of affluence can be seen in Wyoming; 8 counties are within the most affluent 5%, but only one is located within a metropolitan area.

Table 4.2 Metropolitan areas with counties in the most affluent [50] and 5%: 1980

Metropolitan area	Counties	Metropolitan area	Counties
Washington, DC MD VA	[7] 11	Allentown-Bethlehem, PA NJ	1
Chicago-Gary-Lake County, IL IN WI	[4] 10	Amarillo, TX	[1] 1
New York-Northern New Jersey-Long Island, NY NJ CT	[4] 10	Beaumont-Port Arthur, TX	1
San Francisco-Oakland-San Jose, CA	[4] 6	Boston-Lawrence-Salem, MA NH	1
Minneapolis-St Paul, MN WI	[1] 5	Casper, WY	[1] 1
Philadelphia-Wilmington-Trenton PA NJ DE MD	[1] 5	Cedar Rapids, IA	1
Atlanta, GA	[1] 4	Dayton-Springfield, OH	1
Baltimore, MD	[2] 4	Des Moines, IA	1
Detroit-Ann Arbor, MI	[2] 4	Evansville, IN KY	1
Indianapolis, IN	[2] 4	Flint, MI	1
Richmond-Petersburg, VA	4	Fort Wayne, IN	1
Cleveland-Akron-Lorain, OH	3	Kokomo, IN	1
Dallas-Worth Worth, TX	3	Lansing-East Lansing, MI	1
Denver-Boulder, CO	[3] 3	Los Angeles-Anaheim-Riverside, CA	1
Hartford-New Britain-Middletown, CT	3	Midland, TX	1
Houston-Galveston-Brazoria, TX	[1] 3	Norfolk-Virginia Beach-Newport News, VA	1
Kansas City MO-Kansas City, KS	[3] 3	Pittsburgh-Beaver Valley, PA	1
Milwaukee-Racine, WI	[2] 3	Reno, NV	1
St. Louis-East St. Louis-Alton, MO IL	[1] 3	Richland-Kennewick-Pasco, WA	[1] 1
Davenport-Rock Island-Moline, IA IL	2	Roanoke, VA	1
Peoria, IL	2	Rochester, MN	1
Portland-Vancouver, OR WA	2	Rochester, NY	1
Rockford, IL	2	Saginaw-Bay City-Midland, MI	1
Seattle-Tacoma, WA	[1] 2	Steubenville-Weirton, OH WV	1
		Waterloo-Cedar Falls, IA	1

4.2   U.S. Affluence in 1990

The most affluent 5% of counties in 1990 are mapped in Fig. 4.2, and details of their distribution is provided in Table 4.3. In 1980 the most affluent 5% of counties were located in 28 states, and by 1990 this number had expanded to 32. However, these counties were even more concentrated in particular states with 23 in Virginia, 12 in New Jersey, 11 in Maryland, and 9 in both Illinois and New York. The most affluent 50 counties are located in 19 states, with the greatest number in Virginia (9), New Jersey (6), Maryland (5), New York (4), and Connecticut (4). The most affluent county in 1990 was again Los Alamos County, New Mexico, but as mentioned previously the county has a unique history and structure (Encyclopedia Britannica 2021; U.S. Census Bureau 2018a; Wikipedia 2021b).

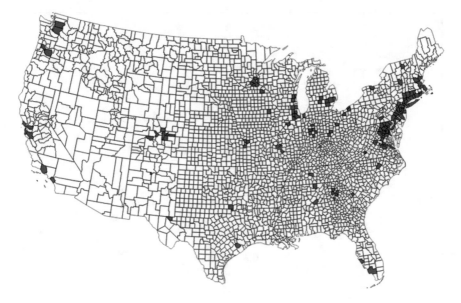

Fig. 4.2 The most affluent 5% of counties, 1990

The most affluent 5% of counties in 1990 was even more concentrated in metropolitan areas in 1990 than they were in 1980. The cluster of rural affluent counties in Wyoming is no longer apparent, and of the 157 most affluent counties 151 are located in 55 metropolitan areas. In 1980 the number of metropolitan areas with affluent counties was 49 so the number of these metropolitan located counties expanded. However, it is clear that affluent counties further concentrated in just a few metropolitan areas that typically cross state boundaries (Table 4.3). Washington DC, Chicago, Philadelphia, and New York have 17, 14, 8, and 8 respectively in the most affluent 5% of counties, and 28 of the 50 most affluent counties are in those same metropolitan areas (Table 4.3). It is relatively easy to visually identify major metropolitan areas on Fig. 4.2, and the 'BosWash' megalopolis stands out as a region of affluence even more so than it did in 1980.

4.3 U.S. Affluence in 2000

The discussion about spatial concentrations of poverty in 1980, 1990, and 2000 point to persistence in certain cores, even as the poorest county changed and exhibited very different characteristics in 2000 than it did in previous decades. This spatial consistency seems to be true of affluent counties also, and in particular regarding their metropolitan nature. Of the most affluent 5% of counties (157) 144 were in 54 metropolitan areas and another 8 were in 8 micropolitan counties. The micropolitan designation was officially added by the Census Bureau in 2003 and thus can be

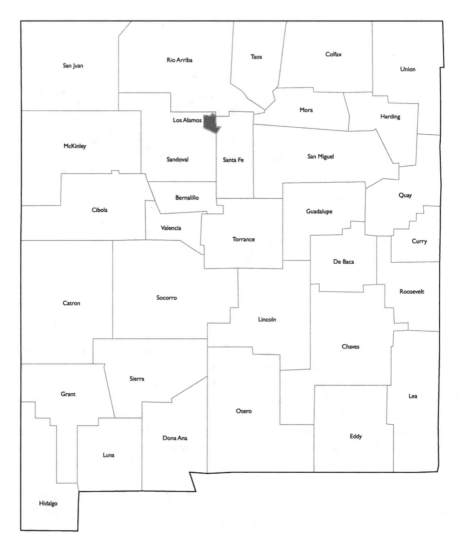

Fig. 4.3 New Mexico counties (Los Alamos)

applied to the 2000 data. A micropolitan statistical area "must have at least one urban cluster of at least 10,000 but less than 50,000 population" (U.S. Census Bureau 2018b). Figure 4.4 again clearly illustrates that large cities tend to be the locations of affluent counties, and the 'BosWash' megalopolis continues to stand out, while Denver seems prominent by 2000.

The distribution of affluent counties in particular metropolitan areas is detailed in Table 4.4. Washington DC and New York clearly dominate with respect to the most affluent 50 counties with 17 and 6 respectively. The other 27 of these most

Table 4.3   Metropolitan areas with counties in the most affluent [50] and 5%: 1990

Metropolitan area	Counties	Metropolitan area	Counties
New York-Northern New Jersey-Long Island, NY NJ CT	[13] 17	Albany-Schenectady-Troy, NY	1
Washington, DC MD VA	[10] 14	Allentown-Bethlehem-Easton, PA NJ	1
Chicago-Gary-Lake County, IL IN WI	[3] 8	Augusta, GA SC	1
Philadelphia-Wilmington-Trenton, PA NJ DE MD	[2] 8	Birmingham, AL	1
Richmond-Petersburg, VA	[2] 7	Boston-Lawrence-Salem, MA NH	1
Atlanta, GA	[3] 6	Burlington, VT	1
Baltimore, MD	[2] 6	Cedar Rapids, IA	1
Indianapolis, IN	[1] 5	Charlotte-Gastonia-Rock Hill, NC SC	1
Minneapolis-St. Paul, MN WI	5	Charlottesville, VA	1
Denver-Boulder, CO	[1] 4	Cincinnati-Hamilton, OH KY IN	1
Detroit-Ann Arbor, MI	[2] 4	Columbus, GA AL	1
San Francisco-Oakland-San Jose, CA	[2] 4	Dayton-Springfield, OH	1
Cleveland-Akron-Lorain, OH	3	Fitchburg-Leominster, MA	[1] 1
Kansas City, MO KS	[1] 3	Hagerstown, MD	[1] 1
Milwaukee-Racine, WI	[1] 3	Harrisburg-Lebanon-Carlisle, PA	1
Norfolk-Virginia Beach-Newport News, VA	3	Houston-Galveston-Brazoria, TX	1
Providence-Pawtucket-Fall River, RI MA	[1] 3	Huntsville, AL	1
St. Louis, MO IL	[1] 3	Kokomo, IN	1
Dallas-Fort Worth, TX	[1] 2	Louisville, KY IN	1
Hartford-New Britain-Middletown, CT	[1] 2	Naples, FL	1
Lansing-East Lansing, MI	2	Nashville, TN	1
Los Angeles-Anaheim-Riverside, CA	2	New London, CT RI	1
Manchester, NH	2	Peoria, IL	1
Portland-Vancouver, OR WA	2	Poughkeepsie, NY	1
Rochester, NY	2	Raleigh-Durham, NC	1
Seattle-Tacoma, WA	2	Roanoke, VA	1
		Santa Fe, NM	[1] 1
		Sarasota, FL	1
		Waterbury, CT	1

affluent counties are spread over 18 metropolitan areas and one micropolitan area, while two are located in non-metropolitan counties. Regionally, it is the 'BosWash' megalopolis that stands out with the Washington DC and New York clusters being added to by multiple affluent counties in Baltimore, Boston, Hartford, Philadelphia, Providence, Richmond, Roanoke, and Virginia Beach.

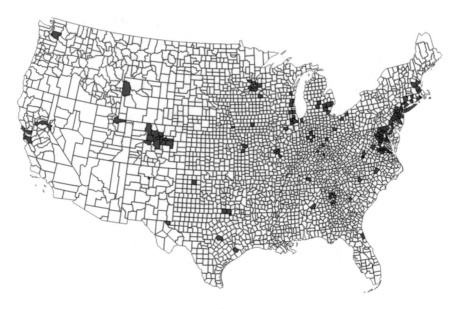

Fig. 4.4 The most affluent 5% of counties, 2000

Table 4.4 Metropolitan and *Micropolitan* areas with counties in the most affluent [50] and 5%: 2000

Metropolitan area	Counties	Metropolitan area	Counties
Washington-Arlington-Alexandria, DC VA MD WV	[11] 17	Allentown-Bethlehem-Easton, PA NJ	1
New York-Newark-Edison, NY NJ PA	[6] 11	Ann Arbor, MI	1
Minneapolis-St. Paul-Bloomington, MN WI	[4] 8	Appleton, WI	1
Chicago-Naperville-Joliet, IL IN WI	[1] 7	Augusta-Richmond County, GA SC	1
Denver-Aurora, CO	[1] 7	Austin-Round Rock, TX	1
Richmond, VA	[2] 7	Birmingham-Hoover, AL	1
Atlanta-Sandy Springs-Marietta, GA	[2] 6	Boulder, CO	1
Baltimore-Towson, MD	[2] 6	Bridgeport-Stamford-Norwalk, CT	1
Philadelphia-Camden-Wilmington, PA NJ DE MD	[2] 5	Charlotte-Gastonia-Concord, NC SC	1
Indianapolis, IN	[1] 4	Charlottesville, VA	1
Boston-Cambridge-Quincy, MA NH	[1] 3	Columbus, OH	[1] 1
Dallas-Fort Worth-Arlington, TX	[1] 3	Des Moines, IA	1
Detroit-Warren-Livonia, MI	[2] 3	Houston-Baytown-Sugar Land, TX	1
Kansas City, MO KS	[2] 3	Jacksonville, FL	1
Milwaukee-Waukesha-West Allis, WI	[2] 3	Kokomo, IN	1
San Francisco-Oakland-Fremont, CA	[1] 3	Lexington-Fayette, KY	1

(continued)

Table 4.4   (continued)

Metropolitan area	Counties	Metropolitan area	Counties
St. Louis, MO IL	3	Louisville, KY IN	[1] 1
Virginia Beach-Norfolk-Newport News, VA NC	3	Manchester-Nashua, NH	1
Cincinnati-Middletown, OH KY IN	2	Monroe, MI	1
Cleveland-Elyria-Mentor, OH	2	Nashville-Davidson-Murfreesboro, TN	[1] 1
Hartford-West Hartford-East Hartford, CT	[2] 2	Norwich-New London, CT	1
Lansing-East Lansing, MI	2	Omaha-Council Bluffs, NE IA	1
Providence-New Bedford-Fall River, RI MA	2	Peoria, IL	1
Roanoke, VA	2	Raleigh-Cary, NC	1
		Rochester, MN	1
Micropolitan Areas	Counties	Sacramento-Arden_Arcade-Roseville, CA	1
Edwards, CO	1	Salt Lake City, UT	1
Gardnerville Ranchos, NV	1	San Jose-Sunnyvale-Sana-Clara, CA	[1] 1
Jackson, WY	1	Seattle-Tacoma, WA	1
Lexington Park, MD	1	Trenton-Ewing, NJ	1
Los Alamos, NM	[1] 1		
Pampa, TX	1		
Silverthorne, CO	1		
Torrington, CT	1		

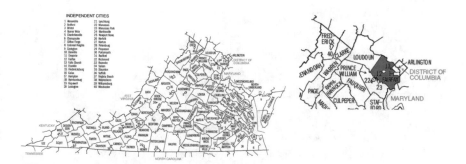

Fig. 4.5 Virginia counties (Falls Church City and Fairfax)

As in both 1980, and 1990, the most affluent county was once again Los Alamos County, New Mexico. Given its unique nature, it is worth noting that the second most affluent county in 1980, 1990, and 2000 was Falls Church City, Virginia, and in both 1990 and 2000 the third most affluent county was Fairfax County, Virginia (Fig. 4.5). Both these areas are part of the 'BosWash' megalopolis which appears to be entrenched as a region of affluence, even when the high cost of living is accounted for by using INDEX as a measure of affluence and poverty.

4.4   U.S. Affluence in 2010

The core areas of affluence in 2010 are displayed in Fig. 4.6, and again a concentration in the 'BosWash' megalopolis is apparent, as well as in multiple metropolitan areas. However, visually on the map, there appears to be more affluent counties west of the Great Plains and outside of the Denver area, than was apparent in 2000 (Fig. 4.6). A change from 2000 is also clear when looking at the makeup of the most affluent 5 percent of counties in 2010 (Table 4.5). Although the list is still dominated by metropolitan counties, there are more micropolitan counties and nonmetropolitan counties in the top 5 percent than in any other of the previous study years. Of the 157 counties 116 were located in 48 metropolitan areas while 13 were located in 12 micropolitan areas. The remaining 28 counties were located in nonmetropolitan areas in Colorado (6), Kansas (2), Massachusetts (1), Montana (2), Nevada (1), North Dakota (5), South Dakota (1), Texas (5), Virginia (2), and Wyoming (3). Clearly the majority of these counties are in the West. While the most affluent counties appear to have dispersed somewhat from the northeast, it is worth noting that the 'BosWash' megalopolis still dominates in terms of the richest of the rich. Of those most affluent 50 counties 30 are in the 'BosWash' megalopolis.

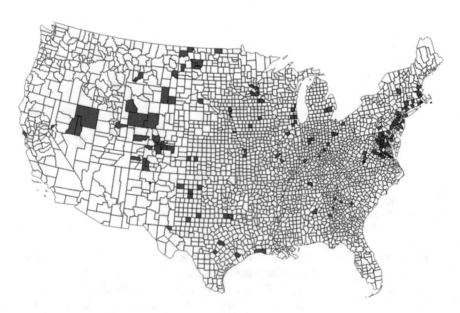

Fig. 4.6 The most affluent 5% of counties, 2010

Table 4.5 Metropolitan and *Micropolitan* areas with counties in the most affluent [50] and 5%: 2010

Metropolitan area	Counties	Metropolitan area	Counties
Washington-Arlington-Alexandria, DC VA MD WV	[12] 15	Albany-Schenectady-Troy, NY	1
New York-Newark-Jersey City-Long Island, NY NJ PA	[5] 7	Athens-Clarke County, GA	1
Denver-Aurora, CO	[1] 6	Augusta-Richmond County, GA	1
Minneapolis-St Paul-Bloomington, MN WI	[2] 6	Austin-Round Rock, TX	1
Richmond, VA	[2] 6	Birmingham-Hoover, AL	1
Baltimore-Towson, MD	[4] 5	Bridgeport-Stamford-Norwalk, CT	1
Philadelphia-Camden-Wilmington, PA NJ DE MD	[1] 5	Cleveland-Elyria-Mentor, OH	1
Chicago-Naperville-Joliet IL IN WI	4	Columbus, GA AL	1
Boston-Cambridge-Quincy, MA NH	[1] 3	Des Moines-West Des Moines, IA	1
Dallas-Fort Worth-Arlington, TX	[1] 3	Evansville, IN KY	1
Indianapolis-Carmel, IN	[1] 3	Houston--Sugar Land-Baytown, TX	1
Virginia Beach-Norfolk-Newport News, VA NC	[3] 3	Lake Charles, LA	1
Amarillo, TX	2	Louisville-Jefferson County, KY IN	1
Atlanta-Sandy Springs-Marietta, GA	[2] 2	Nashville-Davidson-Murfreesboro-Franklin, TN	[1] 1
Charlottesville, VA	2	Omaha-Council Bluffs, NE IA	1
Cincinnati-Middletown, OH KY IN	2	Peoria, IL	1
Detroit-Warren-Livonia, MI	2	Reno-Sparks, NV	1
Hartford-West Hartford-East Hartford, CT	[2] 2	Salt Lake City, UT	1
Kansas City, MO KS	[1] 2	San Antonio, TX	1
Milwaukee-Waukesha-West Allis, WI	[2] 2	San Francisco-Oakland-Freemont, CA	1
Providence-New Bedford-Fall River, RI MA	2	Sioux City, SD	1
Roanoke, VA	2	Sioux Falls, SD	1
Rochester, MN	2	Waterloo-Cedar Falls, IA	1
St. Louis, MO IL	2		
Columbus, OH	[1] 2		
Micropolitan Area	*Counties*	*Micropolitan Area*	*Counties*
Elko, NV	*2*	*Breckenridge, CO*	*1*
		Dickinson, ND	*1*
		Easton, MD	*1*
		Gillette, WY	*[1] 1*
		Glenwood Springs, CO	*1*

(continued)

Table 4.5 (continued)

Metropolitan area	Counties	Metropolitan area	Counties
		Jackson, WY ID	*1*
		Lexington Park, MD	*[1] 1*
		Los Alamos, NM	*[1] 1*
		Rock Springs, WY	*[1] 1*
		Torrington, CT	*1*
		Williston, ND	*1*

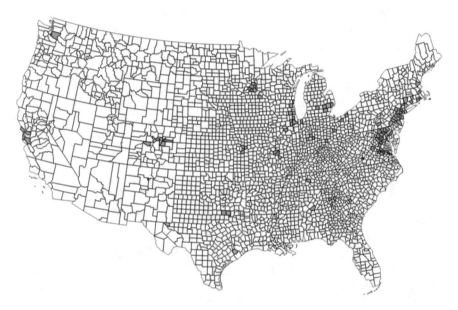

Fig. 4.7 Counties in the most affluent 5% in at least three study years, 1980, 1990, 2000, and 2010

As in previous years Los Alamos County, New Mexico is the most affluent county, followed by Falls Church city, Virginia, and the nonmetropolitan county of Loving, Texas. Loving County borders on New Mexico and is extremely sparsely populated with a total population of 82 in 2010; the county seat Mentone was home to just 19 residents. It is clear from the examples of Los Alamos County and Loving County, that when using counties as the spatial unit of analysis, it is essential to explore the specific characteristics of these areas (Wikipedia. 2021a). In 2010, if these two counties are ignored as being special cases, the most affluent four counties are located in Virginia in the Washington metropolitan area. Thus, it seems that the nation's capital not only represents a node of power, but also of economic privilege.

As with poverty, there seems to be a measure of spatial stability from 1980 to 2010 with regards to the most affluent 5% of counties. Figure 4.7 shows counties

Table 4.6   Counties among the most affluent 50 in at least three years; 1980, 1990, 2000 and 2010

County	In affluent 50 in 1980	In affluent 50 in1990	In affluent 50 in 2000	In affluent 50 in 2010
Marin County, California	Yes	Yes	Yes	No
Douglas County, Colorado	Yes	Yes	Yes	Yes
Middlesex County, Connecticut	No	Yes	Yes	Yes
Tolland County, Connecticut	No	Yes	Yes	Yes
Fayette County, Georgia	Yes	Yes	Yes	Yes
DuPage County, Illinois	Yes	Yes	Yes	No
Hamilton County, Indiana	Yes	Yes	Yes	Yes
Johnson County, Kansas	Yes	Yes	Yes	Yes
Anne Arundel County, Maryland	No	Yes	Yes	Yes
Howard County, Maryland	Yes	Yes	Yes	Yes
Montgomery County, Maryland	Yes	Yes	Yes	Yes
Norfolk County, Massachusetts	No	Yes	Yes	Yes
Oakland County, Michigan	Yes	Yes	Yes	No
Washington County, Minnesota	No	Yes	Yes	Yes
Hunterdon County, New Jersey	No	Yes	Yes	Yes
Morris County, New Jersey	Yes	Yes	Yes	Yes
Somerset County, New Jersey	Yes	Yes	Yes	Yes
Los Alamos County, New Mexico	Yes	Yes	Yes	Yes
Nassau County, New York	Yes	Yes	Yes	No
Putnam County, New York	No	Yes	Yes	Yes
Chester County, Pennsylvania	No	Yes	Yes	Yes
Montgomery County, Pennsylvania	Yes	Yes	Yes	No
Collin County, Texas	No	Yes	Yes	Yes
Loving County, Texas	Yes	No	Yes	Yes
Arlington County, Virginia	Yes	Yes	Yes	Yes
Fairfax County, Virginia	Yes	Yes	Yes	Yes
Hanover County, Virginia	No	Yes	Yes	Yes
Loudoun County, Virginia	No	Yes	Yes	Yes
Alexandria city, Virginia	Yes	Yes	Yes	Yes
Fairfax city, Virginia	Yes	Yes	Yes	Yes

(continued)

Table 4.6 (continued)

County	In affluent 50 in 1980	In affluent 50 in1990	In affluent 50 in 2000	In affluent 50 in 2010
Falls Church city, Virginia	Yes	Yes	Yes	Yes
Ozaukee County, Wisconsin	Yes	No	Yes	Yes
Waukesha County, Wisconsin	Yes	Yes	Yes	Yes

that have been in the most affluent 5% in at least three of the four study years. Looking at the most affluent 50 counties (Table 4.6) reveals that over the study period 15 counties remained in the most affluent 50 in 1980, 1990, 2000, and 2010 and another 18 counties remained in the poorest 50 for three of those years. These 33 counties are located in 18 states, and 9 states were the location of the 15 counties in the most affluent 50 in all four years (Table 4.6). These states are Colorado (1), Georgia (1), Indiana (1), Kansas (1), Maryland (2), New Jersey (2), New Mexico (1), Virginia (5), and Wisconsin (1). There is no state that was the location of counties that stayed in the poorest 50, as well as the most affluent 50, in all four years.

References

Encyclopedia Britannica. 2021. *Los Alamo County, New Mexico, United States*. https://www.britannica.com/place/Los-Alamos-county-New-Mexico Accessed 2 January 2021

U.S. Census Bureau. 2018a. *Quick Faets. Los Alamos County, New Mexico*. https://www.census.gov/quickfacts/losalamoscountynewmexico Accessed 19 February 2019

U.S. Census Bureau. 2018b. *Metropolitan and Micropolitan*. https://www.census.gov/programs-surveys/metro-micro/about.html Accessed 20 February 2019.

Wikipedia. 2021a. *Loving County, Texas*. https://en.wikipedia.org/wiki/Loving_County,_Texas Accessed January 2021.

Wikipedia. 2021b. *Los Alamos County, New Mexico*. https://en.wikipedia.org/wiki/Los_Alamos_County,_New_Mexico Accessed January 2021.

Chapter 5
Changes over the Study Period

Abstract This chapter focuses on the spatial stability and change in terms of poverty and affluence over the study period. The question of relative poverty and affluence is addressed, specifically whether there has been any economic convergence of divergence between the richest and poorest counties. County data for each state is used to explore changes in relative poverty and affluence. Changing inequality in each state is explored using the poorest county INDEX as a percentage of the most affluent county INDEX, and the percentage gap of the richest and poorest counties with respect to the state average. Inequality in 1980, as well as changes from 1980 to 1990, 1990 to 2000, 2000 to 2010, and 1980 to 2010 are discussed. Graphs displaying where each state is positioned in terms of inequality for these periods are provided.

Keywords Spatial stability · Spatial change · Relative poverty · Economic change · Inequality

5.1 Exploring the Poverty – Affluence Gap

It is clear from the previous discussion that there is a measure of spatial stability over the thirty year study period from 1980 to 2010. However, the question about relative poverty and affluence has not been addressed, specifically whether there has been any economic convergence of divergence between the richest and poorest counties.

To get a sense of changes in relative poverty and affluence, county data for each state were used. For each study year two measures were calculated. Firstly, the poorest county INDEX as a percentage of the most affluent county INDEX. Secondly, the percentage gap of the richest and poorest counties with respect to the state average. These two measures provide a view of changing inequality in each state (Table 5.1).

W. Shaw, *America's Poorest and Most Affluent Counties, 1980 to 2010*, SpringerBriefs in Geography, https://doi.org/10.1007/978-3-030-75340-5_5

Table 5.1 The poorest county INDEX as a percentage of that in the most affluent county, and the poorest and most affluent percentage gap compared to the state average in 1980, 1990, 2000, and 2010

State	1980% low of high	1980 low-high % gap	1990% low of high	1990 low-high % gap	2000% low of high	2000 low-high % gap	2010% low of high	2010 low-high % gap
AL	30.9	106.1	25.8	131.1	22.7	144.5	24.5	144.5
AZ	32.1	107.3	24.4	113.3	38.3	88.4	42.1	81.2
AR	31.9	108.6	33.5	99.8	42.0	85.3	31.1	108.0
CA	38.4	105.3	33.2	119.5	31.8	122.6	28.8	126.3
CO	21.8	126.5	23.3	145.0	22.6	151.8	22.1	143.4
CT	75.3	27.0	63.0	31.0	65.6	32.3	69.3	36.9
DE	71.3	35.5	70.0	37.2	73.9	31.2	81.4	21.1
FL	44.3	79.4	39.8	87.4	48.4	73.5	44.3	80.1
GA	27.3	142.3	27.0	153.4	29.1	142.2	21.4	166.5
ID	46.6	72.9	45.9	77.4	42.8	87.8	36.9	93.3
IL	34.1	103.6	31.2	118.4	39.6	93.7	42.4	79.9
IN	46.0	76.6	44.2	86.9	48.2	84.9	43.7	95.0
IA	46.6	72.6	52.5	63.9	57.3	54.4	56.3	57.5
KS	41.1	101.2	41.6	104.8	45.5	92.2	42.6	91.7
KY	28.6	118.7	17.2	156.3	21.8	153.4	19.2	158.9
LA	29.3	114.2	23.0	128.0	37.2	102.1	32.2	112.4
ME	61.7	46.5	63.2	44.6	60.2	48.9	62.8	47.8
MD	38.1	103.1	40.3	91.3	40.4	88.5	35.9	100.1
MA	54.7	59.3	58.8	54.9	55.8	58.6	48.0	68.3
MI	32.2	118.0	31.0	123.0	45.7	87.1	47.8	81.8
MN	37.7	98.9	38.4	100.1	48.9	78.6	44.9	81.7
MS	24.7	126.6	19.9	147.3	25.5	133.3	27.9	130.6
MO	31.8	126.4	30.1	131.4	34.4	116.4	35.6	110.1
MT	51.2	62.8	44.8	81.8	54.7	60.6	45.5	78.9
NE	40.0	84.9	52.6	64.3	46.9	76.3	42.7	80.1
NV	61.7	50.0	58.4	48.4	65.4	42.7	53.1	63.5
NH	75.1	28.7	72.0	33.6	66.1	41.5	61.6	48.9
NJ	51.6	65.3	46.1	76.7	38.6	92.8	31.3	108.8
NM	23.8	188.8	13.9	274.9	16.9	254.7	18.4	249.0
NY	32.3	74.1	29.6	117.2	10.2	150.5	12.9	131.2
NC	41.6	89.7	39.7	94.5	43.3	90.6	38.9	96.0
ND	39.2	92.5	36.3	94.5	33.6	89.6	27.8	107.9
OH	40.9	79.2	41.0	80.2	40.6	95.5	32.9	123.8
OK	36.8	102.7	34.8	101.5	52.3	68.6	42.2	84.7
OR	45.2	74.7	45.7	77.1	55.0	60.9	58.5	55.0
PA	52.9	66.3	41.9	96.5	43.9	93.4	40.2	100.2
RI	79.9	21.9	75.3	26.6	61.4	44.1	55.9	51.2
SC	46.3	74.7	39.8	87.8	41.9	80.0	26.1	106.7

(continued)

Table 5.1   (continued)

State	1980% low of high	1980 low-high % gap	1990% low of high	1990 low-high % gap	2000% low of high	2000 low-high % gap	2010% low of high	2010 low-high % gap
SD	32.8	113.5	22.3	112.2	14.4	127.0	21.4	116.7
TN	26.3	115.7	28.5	127.6	31.6	132.0	25.3	160.4
TX	21.0	137.8	13.4	177.6	14.0	183.9	10.7	245.3
UT	43.8	78.7	37.1	91.2	35.2	106.6	42.0	92.1
VT	69.6	37.8	61.2	50.8	67.4	38.8	67.7	37.9
VA	30.7	137.5	21.3	163.3	26.2	148.1	17.6	190.5
WA	51.2	66.5	53.5	65.9	51.2	69.0	48.4	71.4
WV	37.0	106.0	34.5	105.9	36.3	98.3	36.1	104.1
WI	36.0	105.1	22.6	128.6	36.5	93.8	26.1	116.1
WY	61.1	49.1	57.7	55.8	51.4	67.8	54.2	64.0

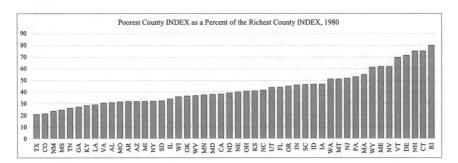

Fig. 5.1 State poorest county INDEX as a percent of the most affluent county INDEX by state, 1980

5.2   The Poverty – Affluence Gap in 1980

In 1980 the level of INDEX in the poorest county expressed as a percentage of that in the most affluent county varied in a range from 21.02 in Texas (Starr County 1382.49 and Loving County 6576.49) to Rhode Island 79.88 (Providence County 3901.90 and Kent County 4884.75) (Fig. 5.1). There appears to be a regional dimension to the inequality between the poorest and most affluent counties at the state level. Of the 10 states with the lowest INDEX expressed as a percentage in the most affluent county, six are in the southeast (AL, GA, KY, LA, MS, TN). At the other end of the spectrum, seven New England states dominate the top 10 showing the least inequality (CT, DE, MA, ME, NH, RI, VT). The percentage gap was highest in New Mexico at 188.83 (the most affluent county was 247.65% of the average while the poorest was 58.82%). The lowest gap was in Rhode Island (the most affluent county was 108.78% of the average while the poorest was 86.89%), resulting in a 21.89 percentage gap (Fig. 5.2). Of course the size of states, including the number of counties, certainly influences these characteristics, and as indicated in earlier

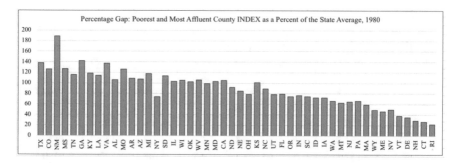

Fig. 5.2 Poorest and most affluent percentage gap compared to state average, 1980

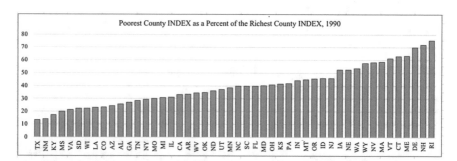

Fig. 5.3 State poorest county INDEX as a percent of the most affluent county INDEX by State, 1990

discussion, context must be considered. Changes over time perhaps provide a general sense of whether convergence or divergence is occurring.

5.3 Changes in the Poverty – Affluence Gap in 1980 to 1990

By 1990 Texas remained the state where the poorest county (Starr) had an INDEX that was lowest compared the most affluent county (Collin), and Rhode Island remained at the other end of the continuum (Fig. 5.3). In Texas the percentage dropped dramatically with the level of INDEX in the poorest county, expressed as a percentage of that in the most affluent county, of only 13.39%. In Rhode Island the change was less dramatic with the percentage dropping somewhat to 75.30. Of the 10 states with the lowest INDEX compared to the most affluent county in 1980, 7 states remained in this group by 1990 – namely Texas, New Mexico, Kentucky, Mississippi, Virginia, Louisiana, and Colorado. Tennessee, Georgia, and Alabama exited this bottom 10, to be replaced by South Dakota, Wisconsin, and Arizona. The other end of the spectrum continued to be dominated by states in New England.

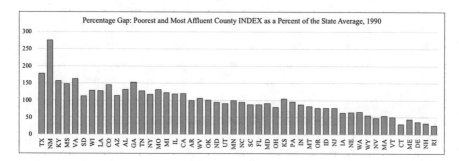

Fig. 5.4   Poorest and Most Affluent County Percentage Gap Compared to State Average, 1990

The percentage gap remained highest in New Mexico where it increased significantly to 274.85 (Fig. 5.4). As was the case in Texas, in New Mexico the poorest county percentage compared to the most affluent county dropped precipitously from 23.75 to 13.9 (Table 5.1). Rhode Island also remained the state with the lowest percentage gap at 26.58.

Looking at the change in percentage gap from 1980 to 1990 New Mexico represents the most extreme increase, followed by New York, Texas, Kentucky, Pennsylvania, Virginia, Alabama, Wisconsin, Mississippi, and Montana (Fig. 5.5). Only 11 states had a reduction in the percentage gap during the 1980s which is an indication of some convergence between the rich and poor counties. These states were Nebraska, Maryland, Arkansas, Iowa, Massachusetts, Maine, Nevada, South Dakota, Oklahoma, Washington, and West Virginia. The magnitude of this reduction in the percentage gap is smaller than the increasing percentage gap in other states, varying between 20.61 in Nebraska and 0.06 in West Virginia. Thus it seems that in general, the geographic expression of inequality grew during the 1980s.

5.4   Changes in the Poverty – Affluence Gap in 1990 to 2000

By 2000 it is clear that there is a lot of geographic stability in how states fare in terms of INDEX in the poorest county as a percentage of that in the most affluent county. Certainly there is some jockeying of position, but New England states are still heavily represented in the top 10 (Fig. 5.6). Perhaps most interestingly, the bottom 10 in 2000 are all states that were in the bottom grouping in either 1980 or 1990, with one notable exception. New York state appears to have the lowest level of INDEX in the poorest county (New York County, $1825), expressed as a percentage of that in the most affluent county (Putnam, $17,926), at 10.18 (Table 5.1). This position is not primarily a result of convergence in states such as Texas and New Mexico, but rather a dramatic divergence in New York from 29.55% in 1990.

The percentage gap in 2000 remains higher in Texas, New Mexico, Kentucky, and Colorado than in New York (Table 5.1; Fig. 5.7). However, Fig. 5.8 reveals that

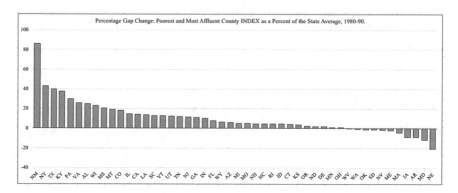

Fig. 5.5 Change in percentage gap between poorest and most affluent counties compared to state average, 1980–90

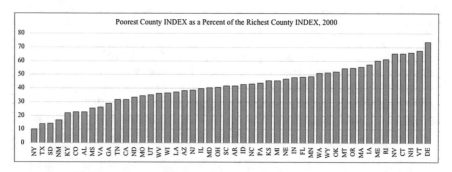

Fig. 5.6 State poorest county INDEX as a percent of the most affluent county INDEX by state, 2000

from 1990 to 2000 a majority of states closed the percentage gap. Twenty nine states narrowed the gap between the poorest and most affluent county, often by a considerable amount. Michigan, Wisconsin, and Oklahoma narrowed the gap by over 30 percentage points, and Louisiana, Arizona, Illinois, Minnesota, Montana, and New Mexico by over 20 percentage points, reversing the general trend in the 1980s. Of the states with an increasing percentage gap New York stands out with over a 33 percentage point increase.

5.5   Changes in the Poverty – Affluence Gap in 2000 to 2010

By 2010 the picture of inequality within states had not changed much. The listing of the bottom 10 remained the same in 2010 as in 2000 except that Tennessee entered the list and Mississippi left it. Within this bottom ten, 5 states appeared in all four study years; Colorado, Kentucky, New Mexico, Texas, and Virginia. Another four

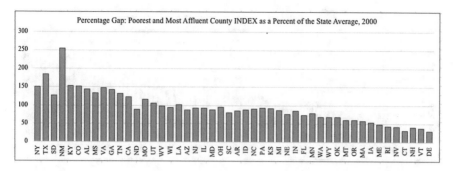

Fig. 5.7 Poorest and most affluent county percentage gap compared to state average, 2000

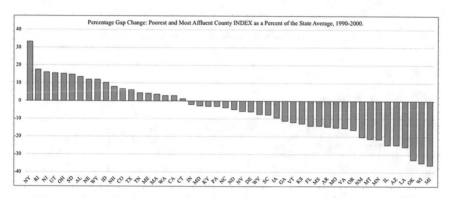

Fig. 5.8 Change in percentage gap between poorest and most affluent counties compared to state average, 1990–2000

appeared on the list in three of four years (Alabama, Georgia, Mississippi, South Dakota), three appeared in two of the years (Louisiana, New York, Tennessee), and just two states placed in the bottom list for a single year (Arizona, Wisconsin) (Table 5.1). In 2010 Texas once more became the state where the poorest county INDEX was the lowest percent of that in the most affluent county, but New York fared little better. The percentage in Texas was 10.68 compared to 12.85 in New York. At the other end of the spectrum were Delaware and Connecticut with percentages of 81.37 and 69.25 respectively (Fig. 5.9).

The largest percentage gap in 2010 was in New Mexico (248.97) closely followed by Texas (245.31) (Table 5.1; Fig. 5.10). The narrowest percentage gap was in Delaware (21.08) followed by Connecticut (36.90) and Vermont (37.89). The narrowest gaps continue to be in the New England region.

Looking at the change in percentage gap from 2000 to 2010, the gap widened in 31 states and narrowed in 17. The biggest narrowing of the percentage gap was in New York where the gap shrank by a little over 19 points. In 9 states the percentage gap expanded by more than 19 points, with a maximum in Texas of a 61.43 percentage point increase (Fig. 5.11). It seems that the general narrowing of the gap

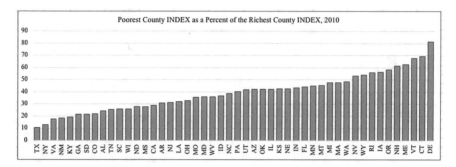

Fig. 5.9 State poorest county INDEX as a percent of the most affluent county INDEX by state, 2010

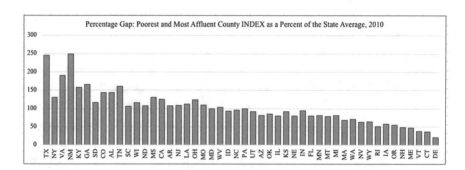

Fig. 5.10 Poorest and most affluent county percentage gap compared to state average, 2010

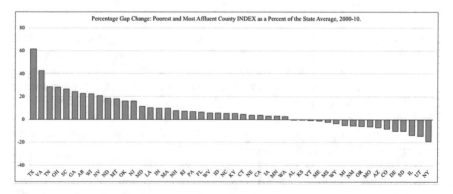

Fig. 5.11 Change in percentage gap between poorest and most affluent counties compared to state average, 2000–2010

observed in the 1990s did not continue in the first decade of the twenty-first century, but instead is similar to the widening gaps seen in the 1980s.

5.6 Changes in the Poverty – Affluence Gap in 1980 to 2010

As discussed previously, a majority of states experienced a widening percentage gap between 1980 and 1990 as well as from 2000–2010, while from 1990 to 2000 the gap narrowed. Figure 5.12 illustrates the combined impact of these changes over the 30-year study period by looking at the change in the gap from 1980 to 2010. Over this thirty year period the percentage gap increased in 33 states and narrowed in 15. Texas showed the largest increase in percentage gap at 107.48, while Michigan's gap shrank the most by 36.23 points. It seems that there has been little progress in geographic inequality over the last thirty years, and in most states such inequality has increased. As can be seem visually in Fig. 5.13, the states where spatial inequality has shrunk to some extent are Michigan, Arizona, Illinois, Oregon, Oklahoma, Minnesota, Missouri, Iowa, Delaware, Kansas, Nebraska, Maryland, West Virginia, Louisiana, and Arkansas. While this listing may seem extensive, 6 of these states saw a minimal single digit contraction in the gap, leaving only 9 states with a double-digit change. In contrast 24 of the 33 states that displayed a widening gap experienced at least a double digit change.

Finally, if the data for all counties in the contiguous United States is explored (Table 5.2), it is clear that spatial inequality in general has been sustained or even strengthened over the 30-year study period. INDEX in the poorest county, expressed as a percentage of that in the most affluent county, dropped significantly from 1980 (13.63) to 1990 (9.03), and dropped again in 2000 (7.10). There was a slight rise in 2010, and the same pattern is evident when looking at the INDEX in the poorest county as a percentage of average. In contrast the percentage of the average in the most affluent county went up consistently in all four study years. In 1980 compared

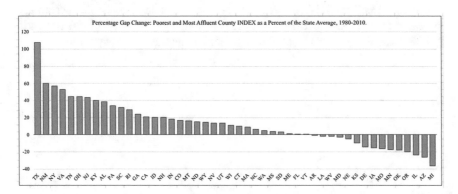

Fig. 5.12 Change in percentage gap between poorest and most affluent counties compared to state average, 1980–2010

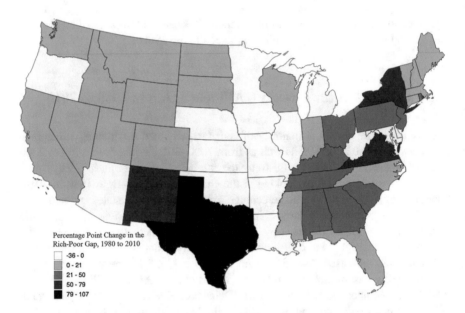

Fig. 5.13 Change in percentage gap between poorest and most affluent counties compared to state average, mapped by state 1980–2010

Table 5.2 All conterminous U.S. counties 1980, 1990, 2000, and 2010

Year	Lowest INDEX	Highest INDEX	% low of high	Lowest % of average	Highest % of average	Low-high % gap	Change in gap
1980	1107.6	8126.1	13.6	29.2	213.9	184.7	
1990	1583.1	17533.2	9.0	24.0	265.4	241.5	56.7
2000	1825.0	25707.4	7.1	18.0	253.3	235.3	−6.2
2010	2683.0	34990.1	7.7	21.7	283.1	261.4	26.1

to 2010 the situation in the poorest county compared to the average got worse dropping from 13.63 to 7.67, whereas in the most affluent county the percentage increased from 213.89 to 283.09 and the percentage gap increased from 184.74 to 261.38. Spatially it seems that the rich got richer and the poor got poorer.

Chapter 6
Characteristics of Poor Versus Affluent Counties

Abstract A statistical analysis of the 1980–1990 data is used to identify the concomitants of income poverty and characteristics of the poverty population. This analysis forms one basis for discussion of representative counties as case studies in each of the cores of poverty as well as examples of affluent counties. In addition, the author discusses some general characteristics of poor versus affluent counties.

Keywords Statistical analysis · Case studies · Generalized characteristics

6.1 A Statistical Analysis of the 1980–1990 Data

In order to provide a baseline for investigating the concomitants of income poverty and characteristics of the poverty population, at the beginning of the period under study, a statistical analysis was conducted on the 1980–1990 data. The literature suggests that income poverty is linked to education, housing, health, occupational structure, residential status, race, age, gender, family structure, migrational tendency, and employment status. Thus, appropriate variables were selected to represent these likely concomitant dimensions of poverty. A correlation analysis of the 25 variables used indicated significant issues of multicollinearity, and so Principal Components Analysis (PCA) was employed to overcome this difficulty. The PCA resulted in twelve factors, which were subsequently used in regression analysis using INDEX as the dependent variable (Shaw 1996).

Based on the PCA factor loadings, which indicate the relationship between the original 25 variables and the twelve factors, the twelve factors were interpreted to represent the following independent dimensions within the data: 1. The black-white duality or racial division; 2. Non-Yuppieness (high young or old population, not highly educated, not urban); 3. Extreme aged population; 4. Size of the urban retail economy; 5. Employment status (unemployment); 6. Manufacturing or the blue-collar economy; 7. Hispanic culture and ethnicity; 8. Health service economy; 9. Native American culture and ethnicity; 10. Health care availability; 11. Population growth; 12. Labor mobility (Shaw 1996).

© The Author(s), under exclusive license to Springer Nature
Switzerland AG 2021
W. Shaw, *America's Poorest and Most Affluent Counties, 1980 to 2010*,
SpringerBriefs in Geography, https://doi.org/10.1007/978-3-030-75340-5_6

Using these twelve factors as independent variables in a regression analysis, with INDEX as the dependent variable, yielded the following results. The direction of the relationship between INDEX and each of the factors reveals that a higher INDEX was negatively related to blackness, non-yuppieness, extreme age population, unemployment, Hispanic culture and ethnicity, and Native American ethnicity, and positively related to the urban retail economy, the health service economy, the manufacturing/blue collar economy, health care availability, population growth, and labor mobility. Four Factors stand out as contributing most to explanation of the magnitude of INDEX. These factors were non-yuppieness, the black-white racial duality, employment status, and the urban retail economy. From this we can conclude that a county with a high percentage of black population, with a high percentage of young or old people, whose population lacks higher education, which has low levels of urbanization and employment, and which lacks an urban retail economy is most likely to have low values of INDEX. A somewhat less substantial but significant influence on INDEX is exerted by the manufacturing or blue collar economy, the health service economy, Hispanic culture and ethnicity, Native American ethnicity, and labor mobility (Shaw 1996).

In sum, these results indicate that race and ethnicity, demographic structure related to age, unemployment and labor mobility, and the structure of the economy, contribute most in understanding the level of INDEX in each county viewing the conterminous United States as a whole. Additional analysis of the 1980–1990 data revealed that these factors were not equally important in the five stable rural cores of poverty that have persisted from 1980 to 2010. In the Western poverty core centered in Arizona-New Mexico the lack of manufacturing activity has a particularly strong influence statistically. The Dakotas poverty core appears to lack urban retail, manufacturing, and health service economies. In the Texas Border poverty core unemployment appears to be a major issue. In the Southern core, poverty is linked to race and its history of slavery and discrimination, as well as unemployment. Finally, the Appalachian poverty core suffers from a lack of diversity within the economy, and the pervasive symptoms of the region's historic and continuing role within capitalism (Shaw 1996). This statistically based study of data relating to the first decade of the study period, can usefully inform case studies of selected counties in the poverty cores, as well as examples of affluent counties.

6.2   County Case Studies in the Poverty Cores

The counties selected as case studies are Apache County Arizona and McKinley County, New Mexico; Buffalo and Todd counties, South Dakota; Maverick and Starr counties, Texas; Greene County, Alabama and Holmes County, Mississippi; Clay and Owsley counties, Kentucky; and the Bronx and Kings County, New York. The selection was based on counties' most consistent position among the poorest counties during the study years in each of the rural poverty cores (Table 6.1).

6.2.1   The Western Core

Apache County Arizona and McKinley County New Mexico have been selected as representative of the Western poverty core. Apache County is located in northeast Arizona and the northern half of the county is part of the Navajo tribal lands (Wikipedia 2020a). The county seat is St. Johns with a population of approximately 3500 (Anon 2020a) out of a total county population of 71,887 (US Census Bureau 2019). Interestingly, the demographic profile in the county seat contrasts with the overall county profile. In St. Johns the population is over 82% White and approximately 10% Native American, compared to 22.9% White and 73.7% Native American for the county as a whole. St Johns is not located within the rugged gorge and canyon dissected Navajo tribal lands in the north, but rather is within the southern portion of the county characterized by grazing lands (Anon 2020a; eReferenceDesk 2020). The first characteristic of Apache County that distinguishes it from national averages is the previously mentioned high percent of Native American population. It addition, the unemployment rate is almost 10% compared to the national average of 3.6. A relatively high percentage of income within the county comes from agriculture, and non-farm jobs are concentrated in the health care and social assistance and the public administration sectors, with very little employment within manufacturing. Net migration in Apache County is negative and population density is low. In addition, a very high percentage of housing units (37.8) are vacant and of all the case study counties, the percentage of households that own a computer is the lowest (Bureau of Labor Statistics 2020; StatsAmerica 2020; US Census 2019).

McKinley County New Mexico forms a contiguous area with Apache County. There are several similarities between the two counties. The population of McKinley County in 2019 was 71,367 of whom 76.7% were Native American. There are two areas of tribal lands within the county, the first a continuation of Navajo Nation land in Apache County, and further south the Zuni Pueblo (Anon 2020b; StatsAmerica 2020; US Census 2019). The county seat is Gallup, which with a population of over

Table 6.1 Poverty rankings of the selected case study counties

County	Poverty rank 1980	Poverty rank 1990	Poverty rank 2000	Poverty rank 2010
Apache, AZ	39	6	29	68
McKinley, NM	77	27	18	72
Buffalo, SD	30	50	2	13
Todd, SD	36	16	13	27
Maverick, TX	16	9	10	18
Starr, TX	3	1	3	3
Greene, AL	5	18	47	23
Holmes MS	4	2	7	10
Clay, KY	11	41	12	17
Owsley, KY	9	3	16	12
Bronx, NY	81	161	4	1
Kings, NY	220	695	31	29

20,000 is a more significant concentration of population than is found in neighboring Apache County. Gallup residents are over 43% Native American, and the city is sometimes called 'the Indian Capital of the World' (Wikipedia 2020b). The economy was once fueled by mining, its function as a railhead, and as a location for the filming of Western movies, but currently a quarter of non-farm jobs are in the area of health care and social assistance with the next largest categories being retail trade and accommodation and food service. As in Apache County, the percentage of employment within manufacturing is relatively low. Also in common with Apache County net migration is negative, and many households do not own a computer (Bureau of Labor Statistics 2020; StatsAmerica 2020; US Census 2019).

6.2.2   The Dakotas Core

Buffalo and Todd counties in South Dakota have been selected as representative of the Dakotas poverty core. Both these counties are home to even higher percentages (Buffalo 85.5 and Todd 82.2%) of Native American population than the counties in the Western core (US Census 2019). Just like the counties in the Western Core, much of the area of Buffalo County consists of tribal lands, in this case the home of Crow Native Americans. The county seat is Gann Valley which is the smallest county seat in the United States with a population of 16 in 2018 (Andrews 2016; Data USA 2020). The total population of Buffalo County is less than 2000, and population density is the lowest of any county selected as a case study. In addition, net migration is negative. Farm earnings are relatively high, and by far the most important sector of non-farm employment is in the area of health and welfare. Another characteristic of life in Buffalo County is the high percentage of residents who rent their housing units. Renting is more common than owner-occupation of housing units, which have a low median value (StatsAmerica 2020; US Census 2019). Todd County, with a population of 10,177, is entirely composed of Lakota tribal lands (Andrews 2015; US Census 2019). The county does not have its own county seat, but instead administration is located in neighboring Tripp County (Wikipedia 2020c). The largest city is Mission with a population of approximately 1200 of the total county population of 10,177. The demographic profile of Mission is similar to that of the county as a whole, with over 76% of residents being Native American (Andrews 2015; Anon 2020c). However in Todd County there is a particularly high percentage of children under 18 years of age, and the official poverty rate of those children is 53%. While there are some jobs in retail trade, in Todd County the largest employment sector is again in the health and welfare category. Just as in Buffalo County more people rent rather than own their homes, and the median value of owner occupied units is lower that in any other county selected as a case study. In both Buffalo and Todd counties the percent of single parent households is the highest of any core (StatsAmerica 2020; US Census 2019).

6.2.3   The Texas Border Core

The counties selected to represent the Texas Border poverty core are Maverick and Starr counties Texas. Maverick County has a total population of 58,722 and the population density of 42.4 per square mile is significantly higher than in either the Western or Dakotas cores. The demographics of the population are also quite different than in the two cores discussed previously. Over 95% of the population in Maverick County identify as Hispanic or Latino and 92.7 are White (StatsAmerica 2020; US Census 2019). The county seat Eagle Pass is relatively large with a population of 28,780, and it is the location of an important border crossing into Piedras Negras, Mexico (Texas Almanac 2020; Wikipedia 2020d). Three sectors of employment stand out in Maverick County; retail trade, accommodation and food services, and public administration. Starr County shares many general characteristics with Maverick County. The residents of Starr County are 95.1% White and 100% Hispanic or Latino. The county is located on the border with Mexico and the county seat Rio Grande City is the location of a border crossing. The population of Starr County is 64,633 and the county seat has a population of approximately 14,000 (StatsAmerica 2020; US Census 2019; Wikipedia 2020e). Employment is Starr County is most concentrated in retail trade and public administration. In Starr County the unemployment rate is high. However, in both Maverick and Starr counties what stands out the most are the very low levels of education and the high percentage of foreign born residents. In Maverick and Starr counties the high school graduation rates are 59.7% and 51.5% respectively. In Maverick County 26.4% of residents aged 25 or older have less than a ninth grade education and in Starr County this percentage is 29.4. In both Maverick and Starr counties the percentage of residents who are foreign born is over 30% (StatsAmerica 2020; US Census 2019).

6.2.4   The Southern Core

The Southern core of poverty is represented by Greene County Alabama and Holmes County Mississippi. Greene County Alabama has a population of approximately 8111 which makes it the least populated county in the state. The county seat and largest city is Eutaw with a population of approximately 2600, but both city and county have lost significant population in recent years (Wikipedia 2020f). The population of Greene County is 79.7% Black, and there is the highest percent of elderly residents aged 65 or older than any other county selected for study in the poverty cores. Greene County stands out as having a significant percentage of jobs in the manufacturing sector as well as in public administration and retail trade. In addition, farm earnings are a significant part of the county economy (StatsAmerica 2020; US Census 2019). Holmes County Mississippi has lost an even greater percentage of its total population than has Greene County from 2010 to 2019. Holmes County is 83.8% Black, and the county seat is the small town of Lexington which is home to

1453 of the total county population of 17,010 (Wikipedia 2020g). Just as in Greene County, Holmes County has its highest percentage of jobs in the manufacturing sector followed by public administration and retail trade. Again, as is the case in Greene County, farm earnings are important to the economy. In Greene County Alabama the percent of housing units that are vacant is very high at over 41% and only 55% of households own a computer. An additional factor that stands out in Holmes County, is the high rate of single parent households (StatsAmerica 2020; US Census 2019).

6.2.5   The Appalachian Core

In both selected Appalachian core counties population loss is an issue. Clay County Kentucky lost 8.4% of its population, and Owsley County Kentucky lost 7.2% between 2010 and 2019. Clay County has a population of 19,901 while Owsley County has a small population (4415) and a lower population density. Clay and Owsley counties are adjacent to each other and share several characteristics. The county seats of both counties are small. Manchester is the county seat of Clay County and has a population a little under 1300, while the county seat of Owsley County is Booneville with a population of just 158 (Wikipedia 2020h, i). Both counties are overwhelmingly White, non-Hispanic Latino, and both have low levels of educational attainment. In Clay County 64.3% of adults aged 25 or older have a high school diploma and 17.6% did not complete the ninth grade; in Owsley County these percentages are 69.1 and 19.1 respectively. Despite these similarities the jobs profile in Clay County versus Owsley County is quite different. Clay County does have jobs in the retail sector, in public administration, and in manufacturing. In contrast, while there are jobs within public administration in Owsley County it is jobs in health care and social welfare that dominate, and farm earnings are important (StatsAmerica 2020; US Census 2019).

6.2.6   The New York City Core

It is clear from the data that the rise of New York City urban counties into the group of most intensively impoverished counties occurred after 1990. The Bronx and Kings County were selected as representative of this urban poverty core, and in many ways they present a very different profile than those found in the five rural cores. Immediately it is clear that both the total populations and the population densities are orders of magnitude greater than in any of the rural cores. The Bronx has a total population over 1.4 million and a population density close to 33 thousand per square mile, while the population of Kings County is approaching 2.6 million and the population density is over 35 thousand per square mile. Unemployment rates in this core are relatively low compared to the other poverty cores, and jobs are distributed across multiple sectors of employment. In both the Bronx and Kings County over one third of jobs are in the area of health care and social assistance. However, in both counties there is also a significant percentage of jobs in retail trade, construction, real estate, educational services, accommodation and food services, and public administration. In the most recent year for which data are available, both the Bronx and Kings County have lost significant population through domestic migration, but have also been the destination for international migrants. Both counties have a diverse population in terms of race and ethnicity, but there are significant differences. In both the Bronx and Kings County over 32% of the population are Black, but in Kings County over 11% are Asian, over 80% of residents are Hispanic Latino, and 43.5% of residents are White. In the Bronx the percent of the White population is half that in Kings County, at almost 56% the percent of the population who identify as Hispanic or Latino is lower, and the percent of Asian population is low. In the Bronx the median income is very low considering the high rents, and over three quarters of residents do rent their accommodations. In Kings County the median salary is higher than in the Bronx, as is the median value of owner occupied housing units, but rents do not differ as substantially. While this urban core contrasts in many ways with the rural poverty cores, there are some similarities to point out. First, there is a link between race and ethnicity and poverty status. In particular, there is a high percentage of Hispanic population just as there is the Texas border core. Also in common with the Texas core, there is a high percentage of foreign born residents, and rates of educational attainment are low (Bureau of Labor Statistics 2020; StatsAmerica 2020; US Census 2019).

6.3 County Case Studies in Affluent Metropolitan Areas

Four counties were selected as case studies representations of the most affluent counties in the United States. Douglas County, CO is in the western United States and is part of the Denver-Aurora-Lakewood metropolitan area; Hamilton County, IN is located in the Midwest and is part of the Indianapolis-Carmel-Anderson metropolitan area; Howard County, MD and Arlington County VA, are both located in the 'BosWash' megalopolis (Table 6.2).

Douglas County, CO is located to the south of central areas of Denver and has a 2019 population of 351,154; the county seat is Castle Rock, population approximately 65,000 (Wikipedia 2020j, k). Hamilton County, IN is located to the north of central Indianapolis and the county seat is Noblesville (Wikipedia 2020l). The population of Hamilton County is very similar to that of Douglas County as is the population of the county seat, however given the smaller area of Hamilton County population density is higher. The first characteristic that stands out in Douglas County is the 23% population growth from 2010 to 2019, and this growth is slightly higher in Hamilton County. These population growth rates are by far the highest of any case study counties. Unsurprisingly, in Douglas County per capita personal income and median household income greatly exceed national averages, and the

Table 6.2 Affluence rankings of the selected case study counties

County	Affluence rank 1980	Affluence rank 1990	Affluence rank 2000	Affluence rank 2010
Douglas, CO	16	15	4	9
Hamilton, IN	27	23	10	20
Howard, MD	7	5	8	7
Arlington, VA	6	9	11	4

official poverty rate (2.6% in 2018) is the lowest of any case study county. Douglas County has 98% of adults ages 25 or older who have a high school diploma, which is the highest rate of any case study county, and the percentage of adults with less than a ninth grade education is only .7 which is the lowest of any case study county. Hamilton County also has high per capita personal income as well as high school graduation rates, but median household income is lower than in Douglas County. However, housing costs in Hamilton County are clearly much lower than in Douglas County. In Douglas County Owner occupation of housing units is the highest, while rentals and vacant units are the lowest of any case study county, but median house value is $441,100 compared to $249,400 in Hamilton County. Another indication of affluence is that these two counties have the highest percentages of computer ownership of any case study counties. Both Douglas and Hamilton counties clearly have diverse economies, with several sectors providing a significant percentage of jobs. In Douglas County the job sectors that stand out are: Retail trade 14.2%, health care and social assistance 10.6%, professional, scientific and technical services 10.3%, accommodation and food services 10%, finance and insurance 9.8%, educational services 8.7%, construction 7.1%. In Hamilton County jobs are concentrated in the following sectors: Retail trade 11.7%, health care and social assistance 11.6%, accommodation and food services 11.3%, finance and insurance 10.4%, professional, scientific and technical services 8.1%, construction 5.5%. The demographic profiles of Douglas and Hamilton counties do not exhibit racial and ethnic diversity compared to national averages except in terms of Asian population. Douglas County is over 89% White, 4.5% Black, and 8.5% Hispanic or Latino, while Hamilton County is 87.1% White, 3.7% Black, and 3.9% Hispanic or Latino. In addition, the number of foreign born residents in both Douglas and Hamilton counties is very low compared to other urban case study counties (StatsAmerica 2020; US Census 2019).

Howard County, MD and Arlington County, VA are both part of the Washington-Baltimore-Arlington Combined Statistical Area (CSA). Howard County is to the southwest of Baltimore city and Arlington County is just west of Washington D.C. across the Potomac River. The population of Howard County is over 325,000 and the county seat is Ellicott City which has a population of over 72,000. Arlington has a population in excess of 236,000 and as a self-governing county, it is its own county seat (Wikipedia 2020m, n). While population growth is not as high as in Douglas and Hamilton counties, growth is still substantial and population densities are quite high. Per capita personal income and median household income are high, but in Arlington County they are higher than any other case study county. Higher income in Arlington County is offset by particularly high housing costs; the median house price is $669,400. In both Howard and Arlington counties not only are high school graduations over 94%, but in Howard County 61.4% of adults aged 25 or older have a bachelor's degree, and in Arlington County that percentage is even higher at 74.6%. In both counties, unemployment rates are low, jobs are within multiple sectors, and there is a particular concentration in the professional, scientific and technical services sector. In Howard County the dominant employment sectors are professional, scientific and technical services 18.1%, health care and social assistance 9.7%, retail trade 9.3%, accommodation and food services 7.2%, and construction 6.6%. In Arlington County 25.6% of jobs are in the professional, scientific and technical services with the other important employment sectors being public administration 15.6%, accommodation and food services 8.6%, health care and social assistance 6.3%, transportation and warehousing 5.6%, and retail trade 5.3%. Howard and Arlington counties are much more racially and ethnically diverse than are Douglas and Hamilton counties. Howard County is 17.7% Asian, 18.7% Black, 58% White, and 6.7% Hispanic or Latino. Arlington County is 10.4% Asian, 9% Black, 71.4% White, and 15.6% Hispanic or Latino. In addition, the population in Howard and Arlington counties are 21.1 and 23.6% foreign born. Finally, it is noticeably that the rate of owner occupied housing units is much lower in Arlington County compared to the other affluent counties profiled, which is likely the result of very high house prices (Bureau of Labor Statistics 2020; StatsAmerica 2020; US Census 2019).

6.4   Some Characteristics of Poor Versus Affluent Counties

As has been discussed previously, poverty tends to be associated with rural countries, while affluence is to be observed primarily in metropolitan areas. Comparing these rural counties with affluent counties can be informative. However, the poorest and the richest counties were both urban in 2000 and 2010, so the question should also be asked how these poor versus affluent urban counties differ from each other in fundamental terms.

6.4.1   Poor Rural Counties Compared to Affluent Counties

The first contrast that stands out is that affluent counties have much larger populations and population densities than do poor rural counties. In addition affluent counties have double digit percentage population growth rates, while three rural poverty cores are losing population and the other two have relatively low growth rates. Income measures provide a stark contrast. The highest 2019 per capita personal income and 2018 median household income in a poor rural case study county are $35,189 (Apache) and $35,954 (Maverick) respectively. In comparison the lowest per capita personal income and median household income in an affluent case study county are $77,263 (Hamilton) and $101,740 (Hamilton). Poverty rates in 2018 in poor rural counties range from 25.9% to 48.4% compared to the range of 2.6% to 6.3% in affluent case study counties. In affluent counties educational attainment is ubiquitously higher than in poor rural counties. In affluent counties high school graduation rates are over 94%, while in poor rural counties the highest rate is 79.6% and dips as low as 51.5%. As has been detailed, jobs in poor rural counties tend to be concentrated in just a few employment sectors, while in affluent counties multiple sectors contribute to a landscape of diverse employment opportunities. In addition, unemployment rates in poor counties are well above the national average, while rates in affluent counties are below this average for the US as a whole. The area of housing provides some additional contrasts. In poor rural counties both house values and owner occupation rates are low, while the percent of vacant housing units is high. With the exception of Arlington County where the home ownership rate is relatively low, in affluent counties a majority of people are home owners, and in all affluent counties house values are high, and the percentage of units that are vacant is low. Access to technology, as evidenced by computer ownership is very different in poor rural case study counties than in affluent counties. Over 95% of households in affluent counties own a computer compared to between 55% and 76.1% in poor rural counties. A final dimension of rural poverty versus metropolitan affluence centers on race and ethnicity. It is striking that in each of the five rural poverty cores one racial or ethnic group dominates the population. In the Western and Dakotas cores the majority of the population are Native American, in the Texas Border core an overwhelming percentage of residents identify as White Hispanic or Latino, in the Southern core over 70% of the population are Black, and in the Appalachian core over 92% of the population are White non-Hispanic Latino. In the selected affluent counties, the two that are located within the 'BosWash' megalopolis have populations that are racially and ethnically diverse, while in the western and midwestern affluent counties the population is predominantly White non-Hispanic Latino. One final demographic racial that stands out as a contrast between poor rural versus affluent counties is the percentage of Asian population which is much higher in affluent counties (Bureau of Labor Statistics 2020; StatsAmerica 2020; US Census 2019).

6.4.2 Poor Urban Counties Compared to Affluent Counties

The Bronx NY, Kings County NY, Howard County MD, and Arlington County VA are representative of U.S. poverty and affluence located in the 'BosWash' megalopolis.

Table 6.3 details some characteristics of these counties in order to explore basic differences between urban counties that represent the richest and the poorest in the US. Even though these affluent and poor counties are all urban, there are some very clear differences evident within the data (U.S. Census Bureau 2019).

First is the contrast in population densities, with the Bronx and Kings counties being much more densely populated than Howard and Arlington counties. This contrast is indicative of the central city location of the poor New York counties, compared to the more suburban location of the two affluent counties. From 2010 to 2019 the population growth rate in Howard and Arlington counties was much higher than in the Bronx and Kings. In the most recent year for which migration data are available, the Bronx and Kings lost population due to domestic out-migration, but this loss was somewhat tempered by significant international in-migration. By comparison Howard County gained some population during that same year but Arlington lost population due to domestic out-migration. Not only are the levels of international in-migration higher in the Bronx and Kings, but so too are the percentages of foreign born residents. The demographic profiles of the Bronx and Kings contrast with those of Howard and Arlington counties. In Howard and Arlington counties the majority of the population is White, while in the Bronx and Kings there is no White majority, over 30% of the population are Black, and a majority of residents identify

Table 6.3   Selected characteristics of the Bronx, Kings, Howard, and Arlington counties

Characteristic	Bronx, NY	Kings, NY	Howard, MD	Arlington, VA
Population 2019	1,418,207	2,559,903	325,690	236,842
Population per square mile (2010)	32,903.6	35,369.1	1144.9	7993.6
Population growth or decline 2010 to 2019	2.4%	2.2%	13.4%	14.1%
Net domestic migration (1 year to 2019)	−31,203	−45,945	710	−2110
Net international migration (1 year to 2019)	8697	8203	779	1057
Asian (2018)	3.6%	11.8%	17.7%	10.40%
Black (2018)	34.1%	32.6%	18.7%	9.0%
White (2018)	21.3%	43.5%	58%	71.4%
Hispanic or Latino % (2018)	55.9%	80.8%	6.7%	15.6%
Foreign born persons, 2014–2018	35.4%	36.5%	21.1%	23.6%
Owner occupied housing units % (2018)	18.6%	27.5%	70%	40.2%
Median value of owner-occupied housing units, 2014–2018	$382,900	$665,300	$448,000	$669,400
Less than 9th Grade (2018)	14.2%	9.2%	2.1%	3.9%
High school diploma or more – Pct. of adults 25+ (2018)	72%	81.6%	95.5%	94.1%
Bachelor's degree or more – Pct. of adults 25+ (2018)	19.8%	36.5%	61.4%	74.6%
Unemployment Rate (2019)	5.4%	4.1%	2.7%	1.9%
Prof, Scientific, and Technical Services Jobs (2019)	1.6%	3.1%	18.1%	25.60%
Health Care, Social Assistance Jobs (2019)	33.1%	33.7%	9.7%	6.3%
Median Household Income (2018)	$38,566	$60,862	$116,719	$120,950
Per Capita Income (2019)	$39,711	$56,080	$79,253	$99,407
Persons in Poverty (by Census Definition, 2018)	27.3%	18.9%	5.2%	6.3%

as Hispanic or Latino. The difference in educational achievement, reflected in both high school graduation rates and college education, is apparent. High school graduation rates in Howard and Arlington counties are 95.5% and 94.1% compared to 72% and 81.6% in the Bronx and Kings. The contrast in the population holding a bachelor's degree is even more stark. In the Bronx and Kings these rates are 19.8% and 36.5% compared to the very high rates of 61.4% and 74.6% in Howard and Arlington counties. A final contrast that is evident regards the dominant sector of employment. In the selected affluent counties, jobs in the professional, scientific and technical sector dominate, but in the poor urban counties it is jobs within health care and social assistance that are the most important and there are few professional, scientific and technical jobs (StatsAmerica 2020; US Census 2019).

References

Andrews, John. 2015. *Land of the Burnt Thigh. South Dakota Magazine*. https://www.southdakota-magazine.com/todd-county Accessed November 2020.

Andrews, John. 2016. *County of Extremes. South Dakota Magazine*. https://www.southdakota-magazine.com/buffalo-county Accessed November 2020.

Anon. 2020a. *World Population Review: St Johns AZ*. https://worldpopulationreview.com/us-cities/st-johns-az-population Accessed November 2020.

Anon. 2020b. *New Mexico Counties: McKinley County*. https://www.nmcounties.org/counties/mckinley-county/ Accessed November 2020.

Anon. 2020c. *World Population Review: Mission SD*. https://worldpopulationreview.com/us-cities/mission-sd-population Accessed November 2020.

Bureau of Labor Statistics. 2020. *Employment by Major Industry Sector*. https://www.bls.gov/emp/tables/employment-by-major-industry-sector.htm Accessed November 2020.

Data USA. 2020. *Gann Valley, SD*. https://datausa.io/profile/geo/gann-valley-sd Accessed November 2020.

eReferenceDesk. 2020. *Apache County, Arizona* https://www.ereferencedesk.com/resources/counties/arizona/apache.html Accessed November 2020.

Shaw, W. 1996. *The Geography of United States Poverty. Patterns of Deprivation, 1980–1990*. Garland Publishing Inc., New York, Fall.

StatsAmerica. 2020. *US Counties in Profile*. https://www.statsamerica.org/USCP/ Accessed November 2020

Texas Almanac. 2020. *Maverick County*. https://texasalmanac.com/topics/government/maverick-county Accessed November 2020.

U.S. Census Bureau. 2019. *Quick Facts*. https://www.census.gov/quickfacts/fact/table/US/PST045219 Accessed November 2020.

Wikipedia. 2020a. *Apache County, Arizona*. https://en.wikipedia.org/wiki/Apache_County,_Arizona Accessed November 2020.

Wikipedia. 2020b. *McKinley County, New Mexico*. https://en.wikipedia.org/wiki/Gallup,_New_Mexico#Demographics Accessed November 2020.

Wikipedia. 2020c. *Todd County, South Dakota*. https://en.wikipedia.org/wiki/Todd_County,_South_Dakota Accessed November 2020.

Wikipedia. 2020d. *Eagle Pass, Texas*. https://en.wikipedia.org/wiki/Eagle_Pass,_Texas Accessed November 2020.

Wikipedia. 2020e. *Rio Grande City, Texas*. https://en.wikipedia.org/wiki/Rio_Grande_City,_Texas Accessed November 2020.

Wikipedia. 2020f. *Greene County, Alabama.* https://en.wikipedia.org/wiki/Greene_County,_ Alabama Accessed November 2020.

Wikipedia. 2020g. *Holmes County, Mississippi.* https://en.wikipedia.org/wiki/Holmes_County,_ Mississippi Accessed November 2020.

Wikipedia. 2020h. *Clay County, Kentucky.* https://en.wikipedia.org/wiki/Clay_County,_Kentucky Accessed November 2020.

Wikipedia. 2020i. *Owsley County, Kentucky.* https://en.wikipedia.org/wiki/Owsley_County,_ Kentucky Accessed November 2020.

Wikipedia. 2020j. *Douglas County, Colorado.* https://en.wikipedia.org/wiki/Douglas_County,_ Colorado Accessed December 2020.

Wikipedia. 2020k. *Castle Rock, Colorado.* https://en.wikipedia.org/wiki/Castle_Rock,_Colorado Accessed December 2020.

Wikipedia. 2020l. *Hamilton County, Indiana.* https://en.wikipedia.org/wiki/Hamilton_County,_ Indiana Accessed December 2020.

Wikipedia. 2020m. *Howard County, Maryland.* https://en.wikipedia.org/wiki/Howard_County,_ Maryland Accessed December 2020.

Wikipedia. 2020n. *Arlington County, Virginia.* https://en.wikipedia.org/wiki/Arlington_County,_ Virginia Accessed December 2020.

Chapter 7
Summary and Conclusions

Abstract This chapter summarizes elements of change and stability in the spatial distribution of both US poverty and affluence. Conclusions and some policy considerations regarding the poverty cores identified are presented.

Keywords Change · Stability · Summary · Conclusions · Policy implications

Mapping INDEX for 1980 and 1990 supports the view that the spatial distribution of American poverty remained essentially unchanged during the 1980s. The five distinct major cores of rural poverty identified to have existed in 1980 clearly persisted until 1990. However, the spatial extent of the cores underwent some change. The poverty area in Dakotas appeared to have contracted just a little. This contraction is not typical of the other cores of poverty. The Western poverty core remained as extensive as in 1980, and the Texas Border core not only persisted but also expanded and became linked by a band of poverty to the Southern poverty region. The Southern poverty region also expanded, especially at its southern extreme in Louisiana. In 1990 the north-south axis of this poverty core stretched along both sides of the Mississippi River from central Arkansas to the Gulf coast. The east-west axis spanned the southern coastal plain to South Carolina to the east, and to the west expanded to include parts of east Texas that had been identified as poor in 1959 by Brunn and Wheeler (1971). The node of intense poverty at the conjunction of the two axes that form this core was still apparent in 1990. Finally, there was also an expansion of the Appalachian core as its southern extent encompassed a large area of West Virginia. Thus it seems that United States poverty became increasingly spatially concentrated from 1980 to 1990 and poverty cores also become more spatially extensive. The five poverty cores identified to exist in 1980 persisted through the decade of the 1980s. Only one of the five poverty regions indicated any sign of contraction while the other four expanded.

Using 2000 and 2010 census information reinforces the notion of extremely stable regions of poverty. While some flux does occur, and some counties such as Tunica County, Mississippi appeared to make substantial progress out of poverty, the Western, Dakotas, Texas Border, Southern, and Appalachian cores remained readily apparent in both 2000 and 2010. Focusing on 2000 these clusters have remained within the poorest of the poor for over twenty years. Only 6 of the 88 counties that were in the poorest 5% in 1980, 1990, and 2000 lie outside these five cores. Of these outliers one is in Arkansas, one in Missouri, one in South Carolina, two in West Virginia, and one in New York. These spatially stable poorest of the poor counties are essentially rural with the exception of the Bronx, New York and Hidalgo and Webb Counties, Texas.

However, the geography of poverty in 2000 compared to the decades previously does indicate some changes. In both 1980 and 1990 only one New York County, the Bronx, appeared among the poorest 5% of counties. By 2000 the Bronx had been joined by two more New York Counties (Kings and New York). Together these three New York counties form a cluster that is difficult to see on the national map because of the small area involved, but this cluster is intensely urban and is home to a large number of people. Of the almost 9 million residents of the most impoverished 156 U.S. counties in 2000, over 5 million (55.8%) lived in these three New York counties. While the other 153 counties are certainly much more visible when mapped, and are important for their spatial extent, they are less significant in terms of the population impacted. Thus, it seems that at the turn of the twentieth century it is urban poverty that is most significant even when poverty is viewed at the county level, a spatial scale that in the past has directed attention to rural poverty regions.

By 2010 it is clear that the five rural cores of poverty that were identified to exist in 1980, have persisted over the 30-year study period. The urban New York cluster of impoverished counties was reduced to two (Bronx and Kings) with New York County no longer appearing in the poorest 5%. However, the Bronx emerged as the poorest county in the conterminous United States in 2010. While poverty remains a largely rural condition when viewed from the spatial scale of the county, experienced in five distinct geographic cores, it seems deeply impoverished urban counties have also emerged beginning in 2000.

This study of poverty at the county level highlights two main points with respect to application and policy. Clearly there are rural regions within the United States where poverty is endemic. These regions are frequently home to minority populations and policy to address impoverishment is these areas must address their very particular cultural, social, and economic landscapes. However, even if poverty in these poor regions were eliminated it would have a limited impact. It is urban counties, such as the Bronx, that are densely populated where large numbers of impoverished people live. In these areas it is cost of living, specifically housing costs, which must be addressed in order alleviate poverty for the substantial number of Americans who live in urban areas.

Over the 30-year study period affluent counties may also be characterized as quite spatially stable. In 1980, while not as spatially clustered as poor counties, they were clearly associated with metropolitan areas. By 1990 affluent counties were even more associated with major metropolitan areas, and the 'BosWash' megalopolis, that crosses multiple state boundaries, stands out as a particular region of affluence. As we entered the first decade of the twenty first century, metropolitan (and micropolitan) counties continued to dominate spatial affluence. Regionally, it is the 'BosWash' megalopolis that stands out, with affluent clusters located in Washington

DC and New York, along with affluent counties in Baltimore, Boston, Hartford, Philadelphia, Providence, Richmond, Roanoke, and Virginia Beach. In addition, Denver seems to have developed as a prominent core of affluence west of the Mississippi.

By 2010 the concentration of affluence in the 'BosWash' megalopolis remains apparent, along with counties associated within multiple other metropolitan areas. There does appear to be an increase in affluent counties west of the Great Plains and in micropolitan counties compared to previous study years.

Case study analysis illustrates structural differences between counties that represent each rural poverty core, the urban New York poverty core, and affluent urban counties. The Western and the Dakotas rural poverty cores both have high percentages of Native-American population many of whom live on tribal lands. In the Dakotas core income is particularly low, official poverty rates are very high, house values are low, the percentage of people who rent their homes is high, and a high percentage of residents are children. These problems are not clearly linked to either unemployment of low levels of education. In the Western core similar issues are apparent, but the depth of multidimensional poverty does not seem to be as intense as in the Dakotas core. It should be noted that far fewer people live the Dakotas core than in the western core, so that while poverty is more intense in the Dakotas far fewer people are impacted.

The Texas Border rural poverty core is notable for its overwhelmingly Hispanic Latino population, over 30% of whom are foreign born. A very high percentage of adults have less than a ninth grade education, and high school graduation rates are exceptionally low. From 2010 to 2019 the population in this core grew significantly, and population density is the highest of any of the rural poverty cores. Very little income is generated through farming, and the biggest employment sector is retail trade. Limited employment opportunities do seem to be a major issue in this poverty core, that must be linked to a fundamental lack of education.

The Southern rural poverty core is also characterized by a distinctive demographic profile, with very high percentage of the population identifying as Black. From 2010 to 2019 this core lost significant population, and a high proportion of residents are either aged under 18 or over 64. It seems likely that working age individuals are leaving the area. It is notable that there is a significant percentage of jobs in the manufacturing, public administration, and retail trade sectors, as well as some farm earnings. The data appear to indicate an area in decline with high vacant housing rates and low rates of owner occupied housing.

The Appalachian core also has its own distinct demographic profile, with 94% of the population identifying as White, non-Hispanic Latino. As in the Southern core there has been significant population loss, and in the most recent year for which data are available, domestic migration, international migration, and natural increase (births minus deaths) are all negative. The percentage of adults who have less than a ninth grade education is quite high and high school graduation rates are relatively low, but in contrast to the Texas Border core very few residents are foreign born.

The urban New York poverty core presents a very different profile than the five rural poverty cores. The most apparent differences are the very large population, high population density, and the ethnic and racial diversity of residents. Also, educational attainment, especially as indicated by the rate adults possess a bachelor's degree, is relatively high and job opportunities are available across multiple sectors of employment. Domestic out-migration is high but international in-migration, natural increase of population, and the percentage of foreign born residents are high. Owner occupation of homes is very low and rental rates are high, likely a result of the very high median value of owner occupied units. These factors combined make it likely that residents who aspire to own their own home are compelled to move away from the center of the city, leaving behind those residents who are not economically able to be mobile. This poverty core has much in common with urban poverty which is typically not visible at the county level, but instead comes into focus using census tracts as the spatial unit (Shaw 2000).

Affluent case study counties appear to be suburban in nature. Their total populations and population densities are not as high as in the poor New York counties, but are higher than in the rural poverty cores. Demographically the 'Boswash' affluent case study counties do exhibit significant ethnic and racial diversity, but it is the percentages of Asian and white, non-Hispanic Latino populations that are high, and there is a significant percentage of foreign born residents. In contrast, both affluent case study counties outside of the 'Boswash' megalopolis do not exhibit racial and ethnic diversity, but instead White, non-Hispanic Latino residents form a very strong majority and there is a much lower percentage of foreign born residents. Across all affluent case study counties, population grew significantly from 2010 to 2019, educational levels are high, unemployment is low, median household income and per capita income are well above the national average, official poverty rates are significantly below the national average, the percentage of elderly population is below the national average, and the vast majority of households have a computer with internet access. Underpinning these shared characteristics across all four affluent case study counties, are multiple employment sectors that provide significant employment. These jobs include those within the professional, scientific, and technical sector which are largely absent from impoverished case study counties.

Focusing on general changes at the state level, over the 30-year study period, a majority of states experienced a widening gap between affluent and poor counties between 1980 and 1990 as well as from 2000 to 2010, while from 1990 to 2000 the gap narrowed somewhat. Over the entire thirty year period using the percentage gap between INDEX for the poorest and most affluent county compared to each state's average INDEX, reveals this gap increased in 33 states and narrowed in 15. Texas showed the largest increase in this percentage gap at 107.48, while Michigan's gap shrank the most by 36.23 points. Thus indications are that there has been little progress in geographic inequality over the last thirty years, and in most states such inequality has increased. This conclusion is supported by looking at changes applied to all counties in the contiguous United States. If the data for all counties in the contiguous United States is explored, it is clear that spatial inequality in general has been sustained or even strengthened over the 30-year study period. The gap between rich and poor, expressed spatially at the scale of the county, widened between 1980 and 2010. If more evidence is needed of the widening gap, consider that in 1980

INDEX in the poorest US county was 1107.60 which had risen to 2683.03 in 2010; an increase of 142% or 1575.43 dollars. By contrast the most affluent county had an INDEX of 8126.07 in 1980 compared to 34,990.13 in 2010. This is an increase of 26,864.06 dollars or 330.6 percent. The distinctive geographies of both poverty and affluence are stable and increasingly spatially pronounced. The poor and the rich clearly continue to live in two different worlds within the United States, just as they did when Lyndon B. Johnson made his declaration of an unconditional 'War on Poverty' in 1964.

References

Brunn, S.D. and Wheeler, J.O. 1971. Spatial Dimensions of Poverty in the United States. *Geografiska Annaler* 53B (1): 6–15.

Shaw, W. 2000. Illinois: Spatial Scales of Poverty. *The Geographical Bulletin*, 47 (1): 9–22, May.

Index

Printed in the United States
by Baker & Taylor Publisher Services